Lecture Notes in Mathematics

Edited by A. Dold and B. Eckmann

995

Banach Spaces, Harmonic Analysis, and Probability Theory

Proceedings of the Special Year in Analysis, Held at the University of Connecticut 1980–1981

T0222476

Edited by R. C. Blei and S. J. Sidney

Springer-Verlag
Berlin Heidelberg New York Tokyo 1983

Editors

Ron C. Blei
Stuart J. Sidney
Department of Mathematics, University of Connecticut
Storrs, CT 06268, USA

AMS Subject Classifications (1980): 42-02, 42A45, 42A61, 42A99, 43A15, 43A46, 46A20, 46B25, 46B99, 46C99, 46J10, 46J15, 47A30, 47B10, 60G99

ISBN 3-540-12314-8 Springer-Verlag Berlin Heidelberg New York Tokyo
ISBN 0-387-12314-8 Springer-Verlag New York Heidelberg Berlin Tokyo

Printed in Germany

Printing and binding: Beltz Offsetdruck, Hemsbach/Bergstr.
2146/3140-543210

PREFACE

During the 1980-81 academic year the Department of Mathematics at the University of Connecticut had the privilege of being host to a Special Year in Analysis. Visitors joined us for periods ranging from a few days to the entire year. Most of them gave talks or series of talks on their recent research. These lecture notes are, roughly, the written versions of most of these expositions. The year's activities, and therefore these notes, focus on topics in Banach spaces, harmonic analysis, and probability, as well as the interactions between these areas; occasional contributions are only secondarily or tangentially in one or more of these areas, with primary emphases elsewhere in analysis.

The papers range in style from detailed analysis to broad exposition. All contain substantial material unavailable elsewhere at this time. We have done far more typographical than mathematical editing, and the papers may not always be polished. We do believe, however, that they are all of interest.

We would like to thank the many visitors who participated in the Special Year and really did make it live up to the name "Special." In alphabetical order, with present affiliations, they are:

Dale E. Alspach (Oklahoma State University, Stillwater)
Aharon Atzmon (Israel Institute of Technology, Haifa)
Jean Bourgain (Free University of Brussels)
Leonard E. Dor (Tel Aviv University)
Sam W. Drury (McGill University, Montreal)
William B. Johnson (Ohio State University, Columbus)
Sten Kaijser (University of Uppsala)
Thomas W. Körner (Cambridge University)
O. Carruth McGehee (Louisiana State University, Baton Rouge)
Daniel M. Oberlin (Florida State University, Tallahassee)
Edward Odell (University of Texas at Austin)
Anthony G. O'Farrell (St. Patrick's College, Maynooth, Ireland)
Aleksander Pełczyński (Polish Academy of Science, Warsaw)
Gilles Pisier (University of Paris VI)
Haskell P. Rosenthal (University of Texas at Austin)

Brent P. Smith (California Institute of Technology, Pasadena)

Mordecay Zippin (Hebrew University, Jerusalem)

We also wish to express our gratitude to the University of Connecticut and the National Science Foundation for making the Special Year possible by their financial support, and to thank the staff of Springer-Verlag and the editors of the Lecture Notes in Mathematics series for their cooperation in helping to bring this volume into being.

Ron C. Blei

Stuart J. Sidney

CONTENTS

PROJECTIONS ONTO $\mathcal{L}_1$ SUBSPACES OF $L_1(\mu)$

Dale E. Alspach[1] and William B. Johnson[2]

ABSTRACT

If X is a $\mathcal{L}_{1,1+\varepsilon}$ subspace of $L_1(\mu)$ and $\varepsilon > 0$ is sufficiently small, then X is complemented.

I. Introduction

Two of the more interesting problems in the almost isometric theory of L_1 spaces are:

(I) Does there exist $\delta > 0$ so that every $\mathcal{L}_{1,1+\varepsilon}$ space with $\varepsilon \leq \delta$ is isomorphic to an $L_1(\mu)$ space? If yes, does the constant of isomorphism tend to one as $\varepsilon \to 0$?

(II) Does there exist $\delta > 0$ so that every $1 + \varepsilon$-complemented subspace of an $L_1(\mu)$ space with $\varepsilon \leq \delta$ is isomorphic to an $L_1(\nu)$ space? If yes, does the constant of isomorphism tend to one as $\varepsilon \to 0$?

In [6], Zippin gave a positive answer to the first part of (I) for subspaces of ℓ_1. Recently Dor [unpublished] showed by a modification of Zippin's argument that the second question in (I) has an affirmative answer for subspaces of ℓ_1.

As one would expect, Zippin's argument in [6] starts with Dor's result [1] that a subspace E of $L_1(\mu)$ with $d(E, \ell^n) \leq 1 + \varepsilon < \sqrt{2}$ is $d(\varepsilon)$-complemented in $L_1(\mu)$, where $d(\varepsilon) \to 1$ as $\varepsilon \to 0$. Since every $L_1(\mu)$ space is norm one complemented in its bidual, Dor's finite dimensional theorem [1] and a standard compactness argument yield that if $X \subseteq L_1(\mu)$ and $d(X, L_1(\mu)) \leq 1 + \varepsilon$, then X is

---

[1] Department of Mathematics, Oklahoma State University, Stillwater, OK 74078, and Department of Mathematics, University of Connecticut, Storrs, CT 06268. Supported in part by NSF MCS-802510.

[2] Department of Mathematics, Ohio State University, Columbus, OH 43210, and Department of Mathematics, Texas A&M University, College Station, TX 77843. Supported in part by NSF MCS-7903042.

$d(\varepsilon)$-complemented in $L_1(\mu)$. On the other hand, since a L_1 space need not be complemented in its bidual, it is not evident how to derive Zippin's theorem [6] from Dor's result [1] and Pełczyński's theorem [5] that every complemented subspace of ℓ_1 is isomorphic to ℓ_1.

In the present note we reduce problem (I) to problem (II) by showing in Corollary 1 that if $\varepsilon > 0$ is sufficiently small, then every $L_{1,1+\varepsilon}$ subspace of $L_1(\mu)$ is $f(\varepsilon)$-complemented, where $f(\varepsilon) \to 1$ as $\varepsilon \to 0$. Besides linking (I) to (II), this yields an alternative proof of Zippin's result [6]; however, it does not yield Dor's unpublished improvement, because the second question in (II) remains unanswered even for subspaces of ℓ_1.

Corollary 1 is a consequence of Theorem 1, which is a simple result concerning complementation in general Banach spaces. Theorem 1 asserts that if $Y^{**} = Y \oplus Z$ satisfies for some $C > 1$,

$\|y + z\| \geq C \min \{\|y\|, \|z\|\}$ $(y \in Y,\ z \in Z)$, and X is a subspace of Y such that X is Δ-complemented in Y^{**} with $\Delta < C$, then X is $g(\Delta)$-complemented in Y, where $g(\Delta) \to 1$ as $\Delta \to 1$. Of course, $L_1(\mu)$ satisfies the condition on Y with $C = 2$, because $L_1{}^{**}(\mu) = L_1(\mu) \oplus_1 L_1(\nu)$ for some measure ν. ($L_1(\mu)^* \cong L_\infty(\mu) \cong C(K)$, for some compact Hausdorff space X. The projection takes a measure γ on K to the part of γ absolutely continuous with respect to μ.) Dor's theorem [1] guarantees that $L_{1,1+\varepsilon}$ subspaces of $L_1(\mu)$ satisfy the condition on X if $\varepsilon > 0$ is sufficiently small.

We use standard Banach space theory terminology, as may be found in the books of Lindenstrauss and Tzafriri [3], [4].

The authors gratefully acknowledge the hospitality of the Department of Mathematics of the University of Connecticut during the time the research for this paper was conducted.

II. The complementation results

The main step in the proof of Theorem 1 is the following approximation result.

Lemma 1. Suppose that P is a projection from Y^{**} onto Y so that for some $C > 0$ and all $y^{**} \varepsilon Y^{**}$

$$(C) \qquad \|y^{**}\| \geq C \min (\|Py^{**}\|, \|(I-P)y^{**}\|) .$$

Then for each subspace X of Y and each $x^{**} \varepsilon X^{**} \equiv X^{\perp\perp} \subseteq Y^{**}$, we have:

$$d(Px^{**}, X) \equiv \inf_{x \varepsilon X} \|Px^{**} - x\| \leq C^{-1} \|x^{**}\| .$$

Proof. Given $x^{**} \varepsilon X^{**}$, take $y^* \varepsilon Y^* \cap X^{\perp}$ so that $\|y^*\| = 1$ and $\langle y^*, Px^{**}\rangle = d(Px^{**}, X)$. Since $y^* \varepsilon X^{\perp}$, $\langle y^*, x^{**}\rangle = 0$ so $0 = \langle y^*, x^{**}\rangle = \langle y^*, Px^{**}\rangle + \langle y^*, (I-P)x^{**}\rangle = d(Px^{**}, X) + \langle y^*, (I-P)x^{**}\rangle$ and, consequently,

$$\|(I-P)x^{**}\| \geq |\langle y^*, (I-P)x^{**}\rangle| \geq$$

$$\geq d(Px^{**}, X) .$$

Since P satisfies (C), we get that $\|x^{**}\| \geq C\, d(Px^{**}, X)$. □

Lemma 2. Suppose that X is a subspace of Y and $T : Y \to Y$ is an operator such that $T|_X = I_X$ and for some $\alpha < 1$ and all $y \varepsilon Y$,

$$(\alpha) \qquad d(Ty, X) \leq \alpha\|y\| .$$

Then there is a projection R from Y onto X such that $\|R - T^n\| \to 0$ as $n \to \infty$ and $\|R\| \leq \inf_{n \geq 1} \{\|T\|^n + (1-\alpha)^{-1} \alpha^n (1 + \|T\|)\} \leq$

$\leq (1-\alpha)^{-1}(1 + \|T\|)$.

Proof. Fix $\alpha < \beta < 1$. Given $y \in Y$, we can use the inequality (α) recursively to choose $x_1, x_2, \ldots$ in X to satisfy for each $n = 1, 2, \ldots$.

$$\text{(i)} \qquad \left\| x_n + T\left(\sum_{i=1}^{n-1} x_i - T^{n-1}y\right) \right\| \leq \beta \left\| \sum_{i=1}^{n-1} x_i - T^{n-1}y \right\| .$$

Since $T|_X = I_X$, we get by induction that for $n = 1, 2, \cdots$.

$$\text{(ii)} \qquad \left\| \sum_{i=1}^{n} x_i - T^n y \right\| \leq \beta^n \|y\| .$$

From (i) and (ii) we conclude that for each $n = 2, 3, \cdots$,

$$\|x_n\| \leq \left\| \sum_{i=1}^{n} x_i - T^n y \right\| + \|T\| \left\| \sum_{i=1}^{n-1} x_i - T^{n-1} y \right\|$$

$$\leq (\beta^n + \beta^{n-1} \|T\|) \|y\|$$

which is to say that

$$\text{(iii)} \qquad \|x_n\| \leq \beta^{n-1} (\beta + \|T\|) \|y\| \qquad (n = 2, 3, \cdots).$$

Inequalities (ii) and (iii) imply that $(T^n y)_{n=1}^{\infty}$ and $(\sum_{i=1}^{n} x_i)_{n=1}^{\infty}$ both converge to the same element, Ry, of X, and that the convergence is uniform on the unit ball of Y. Since obviously $R|_X = I_X$, it remains only to estimate $\|R\|$.

Continuing with the above notation, we have that for each $n = 1, 2, \cdots$,

$$\left\| \sum_{i=1}^{n} x_i \right\| \leq \left\| \sum_{i=1}^{n} x_i - T^n y \right\| + \|T^n y\|$$

$$\leq (\beta^n + \|T\|^n) \|y\| ,$$

which, together with (iii) yields:

$$\|Ry\| = \|\sum_{i=1}^{\infty} x_i\| \le \inf_n \{\|\sum_{i=1}^{n} x_i\| + \sum_{i=n+1}^{\infty} \|x_i\|\}$$

$$\le \inf_n (\beta^n + \|T\|^n + (\beta + \|T\|) \sum_{i=n+1}^{\infty} \beta^{i-1}) \|y\|$$

$$= \inf_n [\|T\|^n + (1-\beta)^{-1} \beta^n (1 + \|T\|)] \|y\|.$$

Since $\beta > \alpha$ is arbitrary, we get the claimed estimate for $\|R\|$. □

Theorem 1. Let $\frac{1}{2} < \alpha < 1 < C \le 2$ with $\alpha C > 1$. Then there is a constant $K = K(\alpha)$ for which the following is true:

Suppose that P is a projection from Y^{**} onto Y which satisfies (C), X is a subspace of Y, and Q is a projection from Y^{**} onto $X^{\perp\perp} \equiv X^{**}$ so that

$$\|Q\| = 1 + \varepsilon \le \alpha\, C\,.$$

Then there is a projection R from Y onto X so that

$$\|R\| \le 1 + K\varepsilon|\log\varepsilon|\,.$$

Proof. Set $T = PQ|_Y$. By Lemma 1, T satisfies condition (α) in Lemma 2 and hence the existence of the projection R follows from Lemma 2. If in the estimate in Lemma 2 for $\|R\|$ we take $n \sim \delta|\log\varepsilon|$, where $\alpha^{-\delta} = e$, then we get the desired estimate for $\|R\|$.

□

To apply Theorem 1 as stated to L_1 spaces, we need the following well-known proposition. (The implication (1)⇒(3) follows from a compactness argument, see [1] or [2], while the implication (3)⇒(1) is an immediate consequence of the principle of local reflexivity.)

Proposition 1. Let X be a subspace of Y and let $1 \le \lambda < \infty$. The following are equivalent:

(1) For each $\gamma > \lambda$ and all finite dimensional subspaces E of X and F of Y with $E \subseteq F$, there is an operator $T : F \to X$ so that $\|T\| \le \gamma$ and $T|_E = I_E$.

(2) There is a projection S on Y^* so that $(I - S)Y^* = X^\perp$ and $\|S\| \le \lambda$.

(3) There is a projection Q from Y^{**} onto $X^{\perp\perp}$ so that $\|Q\| \le \lambda$.

Corollary 1. Suppose that X is a $\mathcal{L}_{1,1+\varepsilon}$ subspace of $L_1(\mu)$ with $1 + \varepsilon \le 3^{-1/2}2 < 1.16$. Then X is complemented in $L_1(\mu)$. Moreover, there is a constant K so that if $\varepsilon \le .1$, X is $1 + K\varepsilon|\log\varepsilon|$-complemented in $L_1(\mu)$.

Proof. Dor [1] proved (something stronger than) that if X is a $\mathcal{L}_{1,1+\varepsilon}$ subspace of $Y = L_1(\mu)$ with $1 + \varepsilon < \sqrt{2}$, then (1) in Proposition 1 holds with $\lambda = \lambda(\varepsilon) = [2(1+\varepsilon)^{-1}-1]^{-1}$. Now $L_1(\mu)$ satisfies (C) with $C = 2$, where P is the norm one band projection of $L_1(\mu)^{**}$ onto $L_1(\mu)$, and $\lambda(\varepsilon) < 2$ if $1 + \varepsilon < 3^{-1/2}2$, so Theorem 1 yields that X is complemented in $L_1(\mu)$. For $\varepsilon \le .1$, $\lambda(\varepsilon) \le 1 + 3\varepsilon$, so the estimate for the norm of the projection from $L_1(\mu)$ onto X follows from the estimate for $\|R\|$ in Theorem 1. □

Remark. We do not know whether the estimate "$1 + K\varepsilon|\log\varepsilon|$" in Theorem 1 and Corollary 1 can be replaced with "$1 + K\varepsilon$".

By using Dor's lattice version (Dor's thesis [2]) of his result in [1], we obtain in a similar manner:

Corollary 2. Given any $1 \le p < \infty$, there exists $\delta = \delta(p) > 0$ and a constant $K = K(p)$ so that if Y is a p-concave Banach lattice with p-concavity constant one and X is a $\mathcal{L}_{1,1+\varepsilon}$ subspace of Y with $\varepsilon \leq \delta$, then X is $1 + K\varepsilon|\log\varepsilon|$ - complemented in Y.

References

[1] L. Dor, On Projections in L_1, Annals of Math. 102 (1975), 463-474.

[2] L. Dor, On Embedding of L_p-spaces in L_p-spaces, Dissertation, The Ohio State University, 1975.

[3] J. Lindenstrauss and L. Tzafriri, Classical Banach Spaces I, Springer-Verlag, Berlin, 1977.

[4] J. Lindenstrauss and L. Tzafriri, Classical Banach Spaces II, Springer-Verlag, Berlin, 1979.

[5] A. Pełczyński, Projections in certain Banach Spaces, Studia Math 19 (1960), 209-228.

[6] M. Zippin, $\mathcal{L}_1$ subspaces of ℓ_1, Israel J. Math 22 (1975), 110-117.

NEW BANACH SPACE PROPERTIES OF THE DISC ALGEBRA AND H^∞

J. Bourgain(*)

The purpose of this note is to describe some striking similarities from the Banach space point of view between the disc algebra A and the space of continuous functions C, between the space H^∞ of bounded analytic functions on the disc and the space L^∞. We will consider

1. Topological properties of H^∞ and its dual

THEOREM 1: (i) The odd duals of H^∞ are weakly complete.

(ii) The even duals of H^∞ are Grothendieck spaces.

(iii) H^∞ and its duals satisfy the Dunford-Pettis property.

2. Projections and decompositions

THEOREM 2: c_0 is up to isomorphism the only complemented subspace of A with an unconditional basis.

THEOREM 3: (1) Finite dimensional complemented subspaces of H^∞ contain ℓ_n^∞'s of proportional dimension n

(2) ℓ^∞ embeds in any infinite dimensional complemented subspace of H^∞

THEOREM 4: H^∞ only admits unconditional decompositions in the ℓ^∞-sense (cfr. [7]).

3. Absolutely summing operators and local theory

THEOREM 5: Let Y be a Banach space such that any operator from C onto Y is p-summing (for some fixed $p \geq 2$). Then all

---

(*) Department of Mathematics, Vrije Universiteit Brussel
Pleinlaan 2 - F7, 1050 BRUSSELS

operators from A into Y are also p-summing. In partic ular $L(A,\ell^1) = \Pi_2(A,\ell^1)$ and A^* has cotype 2.

THEOREM 6: Any 0-summing operator from A into an arbitrary Banach space is nuclear.

These similarities are somehow surprising, since, for instance, C and A have a basically different finite dimensional structure (see [6]). In this spirit, they appear to be also of interest in the frame of general Banach space theory.

The preceding results depend on more concrete analytical facts which we explain briefly in what follows.

For convenience, we see A as a subspace of $C(\Pi)$ and H^∞ as a subspace of $L^\infty(\Pi)$, thus as functions on Π. For the duality $\langle f,\phi\rangle = \int_\Pi f.\phi\, dm$, H^∞ identifies with the dual of the quotient space L^1/H^1_0, where H^1_0 is the subspace of $L^1(\Pi)$ of functions f s.t. $\hat{f}(n) = 0$ for $n \leq 0$. Denote $q : L^1 \to L^1/H^1_0$ the quotient map and $\sigma : L^1/H^1_0 \rightsquigarrow L^1$ the minimum norm lifting.

PROPOSITION 1: Given $\delta > 0$, there exists $\delta_1 > 0$ and a function $\alpha(n)$ s.t. $\frac{\alpha(n)}{n} \xrightarrow{n\to\infty} 0$ so that the following holds.

If $f_1, f_2, \ldots, f_n$ in $L^1(\Pi)$ are δ-Rademacher ℓ^1, i.e. if

$$\int \|\Sigma\, \varepsilon_k\, c_k\, f_k\|_1\, d\varepsilon \geq \delta\, \Sigma\, |c_k|\ \|f_k\|_1 \quad (c_1,\ldots,c_n \in \mathbb{C})$$

and if

$$\|q(f_k)\| \geq (1-\delta_1)\ \|f_k\|_1 \quad (1 \leq k \leq n)$$

(in particular, if the f_k are minimum norm liftings)

then there are H^∞-functions $\phi_1,\ldots,\phi_n$ and $\psi_1,\ldots,\psi_n$ satisfying the following properties:

(i) $|\phi_k| + |\psi_k| \leq 1 \qquad (1 \leq k \leq n)$

(ii) $\Sigma\, |\phi_k| \leq 1$

(iii) $\Sigma\, |1-\psi_k| \leq \alpha(n)$

(iv) $\langle f_k, \phi_k \rangle = \delta_1\, ||f_k||_1$

Prop. 1 is the main ingredient for the proof of Th's 1, 2, 3 and 4. The system (ψ_k) is used only to obtain Th. 1 (ii) and Th. 2 (ii). The existence of functions (ϕ_k) fulfilling (ii) and (iv) generalizes a result of J. Garnett [2] that harmonically interpolating sequences in the disc are interpolating. In this case the functions f_k are just Poisson kernels. Given now disjointly supported, positive, norm-1 functions on Π, we can approximate them by disjointly supported convex combinations of Poisson kernels (P_{z_j}), where (z_j) is interpolating. However, improving the approximation requires a worse interpolation constant. The functions (ϕ_k) can be obtained using the following

<u>PROPOSITION 2</u>: Given $\delta > 0$, there exists $C_\delta < \infty$ such that if (z_j) is a δ-interpolating sequence in the disc and (S_k) are subsets of N so that each integer j is in at most K sets, then there exists a system (ϕ_k) of H^∞ functions satisfying

(i) $\phi_k(z_j) = 1$ for each $j \in S_k$ and each k

(ii) $\Sigma\, |\phi_k| \leq C_\delta \cdot K$

(iii) $||\phi_k||_\infty \leq B$ for each k, where B is a <u>fixed</u> constant.

Because B must not depend on δ, the ϕ_k cannot be obtained just by summation of the P. Beurling functions associated to the sequence (z_j). The system (ϕ_k) is obtained as the solution of a vector-valued interpolation problem. We use the $\bar{\partial}$-technique and certain interpolation properties of vector-valued H^1-spaces.

To construct the functions (ψ_k), we make use of the following refinement of Havin's lemma [3].

PROPOSITION 3: Given $0 < \tau < 1$, there is a constant $C_\tau < \infty$ such that if (A_i) is a sequence of measurable subsets of Π and (ε_i) a sequence of positive numbers, there exist H^∞ functions f and g satisfying

(i) $|f| + |g| \leq 1$

(ii) $|f(z)-1| < \varepsilon_i$ for $z \in A_i$

(iii) $\|f-\tau\|_2^2 \leq C_\tau \sum \varepsilon_i^{-2} m(A_i)$

(iv) $\|g-(1-\tau)\|_2^2 \leq C_\tau \sum \varepsilon_i^{-2} m(A_i)$.

We show that any Hahn-Banach norm-preserving extension $(H^\infty)^* \rightsquigarrow L^\infty(\Pi)^*$ maps weakly conditionally compact sets Ω on relatively weakly compact sets. Moreover, if Ω is not weakly conditionally compact, there are sequences (x_r) in H^∞ and (x_r^*) in Ω such that $\inf |\langle x_r, x_r^* \rangle| > 0$ and $\sum |x_r|$ is L^∞-bounded. This fact is derived from Prop. 1 using general principles such as local reflexivity and ultra-product representation. This is possible because Prop. 1 deals with a finite number of functions and can be reformulated in Banach space terminology.

Let us now concentrate on Th's. 5 and 6. They both depend on the weak-type property of the Riesz projection. To obtain Th. 5, we combine the $(i_p - \pi_p)$-property $(p > 1)$ for the disc algebra (see [6]) with an interpolation inequality for p-summing norms.

PROPOSITION 4: For $1 < p < \infty$ and $T \in \Pi_p(A,Y)$, one has
$i_p(T) \leq \text{const}\ \frac{p^2}{p-1}\ \pi_p(T)$.

PROPOSITION 5: Assume $p > 1$ and T a p-summing operator on A. Let $q > p$ and θ s.t. $\frac{1}{q'} = \theta + \frac{1-\theta}{p'}$. For all $0 \leq \phi < \theta$, there exists a constant $C = C(p,q,\phi)$ s.t. $\pi_q(T) \leq C \, \|T\|^{\phi} \, \pi_p(T)^{1-\phi}$.

Th. 5 implies that A^* has the Orlicz-property and therefore cotype 2, since A is isomorphic to its c_0-direct sum $(\Sigma A)_{c_0}$. In fact, the Rademacher mean can be characterized as follows:

PROPOSITION 6: For $f_1,\ldots,f_n$ in $L^1(\Pi)$, one has the equivalence

$$\int \|\Sigma \varepsilon_k f_k\|_{L^1/H_0^1} \, d\varepsilon \sim \inf_{h_k \varepsilon H_0^1} \int_{\Pi} (\Sigma |f_k + h_k|^2)^{\frac{1}{2}} \, dm$$

In [4] Th. 6 was shown for the space C and it was conjectured that $\Pi_0(A,\ell^2) = N(A,\ell^2)$. The 0-summing property of T is used as follows:

PROPOSITION 7: For g a function on Π and $\theta \varepsilon \Pi$, denote g_θ the translate $g_\theta(\psi) = g(\theta+\psi)$. Let $0 < p < 1$ and $T \varepsilon \Pi_p(A,Y)$. Then, P_r being the Poisson kernel,

(i) The functions $\xi_r(\theta) = T(R(P_{r,\theta}))$ converge in $L^\alpha_Y(\Pi)$ for $r \to 1$ and $\alpha > 1$.

(ii) The function $F(\theta) = \sup_{r<1} \|\xi_r(\theta)\|$ satisfies a weak-type inequality

$$\|F\|_{1,1} \leq C(p) \, \pi_p(T) .$$

It is shown in [4] that the fact that X^* satisfies the Grothendieck theorem, i.e. $L(X^*,\ell^2) = \Pi_1(X^*,\ell^2)$, does not necessarily mean that X verifies Th. 6. This phenomenon is also possible for X a C_Λ-space. Take for instance

$$\Lambda = \mathbb{Z}_+ \cup \{-2^n \; ; \; n=0,1,2,\ldots\} .$$

It is easily seen that then Th. 5 also holds for C_Λ. However (see [4]), the orthogonal projection $P : C_\Lambda \to L^2_{\{-2^n\}}$ is 0-summing and onto.

In proving that $L(A,\ell^1) = \Pi_2(A,\ell^1)$, we use the algebraic structure of A as well as singular integral properties of the orthogonal projection. It should be noticed that if one restricts to (A,ℓ^1)-multipliers, only the second aspect is used. On the other hand, it is straightforward that the equality $L(B,\ell^1) = \Pi_2(B,\ell^1)$ extends to any closed subalgebra B of $L^\infty(\Pi)$ containing H^∞; this is a consequence of the Douglas property [5].

It does not seem to be known if some of the preceding results go through for polydisc- or ball-algebras, or for spaces of smooth functions. In particular, we don't know if an $(i_p - \pi_p)$-theorem is true.

References

[1] J. Bourgain: preprints.

[2] J. Garnett: Interpolating sequences for bounded harmonic functions, Indiana University Math. J. 21 (1971), 187-192.

[3] V. P. Havin: Weak sequential completeness of the space L^1/H^1_0, Vestnik Leningrad, Univ. 13 / 1973, 77-81.

[4] S. V. Kisliakov: What is needed for a 0-absolutely summing operator to be nuclear, preprint.

[5] D. E. Marshall: Subalgebras of L^∞ containing H^∞, Acta Math., 137 (1976), 91-98.

[6] A. Pełczyński: Banach spaces of analytic functions and absolutely summing operators, Conf. Board of Math., Sci., Regional Conf. Ser. in Math. n° 30 (1976).

[7] P. Wojtaszczyk: Decompositions of H^p spaces, Duke Math. J., Vol. 46, N 3 (1979), 635-644.

REMARKS ON VON NEUMANN'S INEQUALITY

S. W. Drury

This article contains a rather personal exposition of various aspects of Von Neumann's inequality and its generalizations. Included are a number of observations made by this author and doubtless also by others. The purpose of the article is to stimulate interest in understanding the situation for three or more commuting contractions.

§ 1. Introduction

Throughout this article all Hilbert spaces will be complex. The original inequality of von Neumann asserts that if T is a linear contraction on Hilbert space (that is a linear endomorphism with operator norm $\|T\|_{op}$ bounded by one) and p is a complex polynomial satisfying

$$\sup_{|z|\leq 1} |p(z)| \leq 1$$

then $P = p(T)$ the result of substituting T into the polynomial p is again a linear contraction. Indeed it follows that $p(T)$ may be defined for every element p of the disc algebra.

Von Neumann's proof [1] is based on the special case in which p is a Möbius function. Let $\lambda \in C$, $|\lambda| < 1$ and set $p(z) = (z-\lambda)(1-\bar{\lambda}z)^{-1}$. Then an easy calculation shows that

$$I - P^*P = (1-|\lambda|^2)(I-\lambda T^*)^{-1}(I-T^*T)(I-\bar{\lambda}T)^{-1} .$$

Since T is a contraction, $I - T^*T$ is positive definite: the identity shows that $I - P^*P$ is again positive definite and it follows that P is a contraction.

The proof of the inequality now rests on the following result [2].

<u>Theorem</u>. The space B of absolutely convergent sums of finite Blashke products is isometrically isomorphic to the disc algebra.

Since the inequality holds for Möbius functions it holds for finite Blashke products and hence by the theorem for all elements of the disc algebra.

§ 2. <u>Unitary dilations</u>

The inequality is trivial if T is normal. In the language of spectral theory we then have a spectral measure Q defined on the closed unit disc such that

$$T = \int_{|z|\leq 1} z \, dQ(z) .$$

From this

$$p(T) = \int_{|z|\leq 1} p(z) dQ(z)$$

and $p(T)$ is a contraction since $|p(z)| \leq 1$ on the domain of integration.

The proof of Halmos [3], [4] exploits this fact. Given now an arbitrary contraction T on a Hilbert space $\mathcal{H}$ one constructs a Hilbert space $\mathcal{K}$, an isometric inclusion $J: \mathcal{H} \to \mathcal{K}$ and a unitary U on $\mathcal{K}$ such that

$$T^n = J^* U^n J \qquad \forall n \in Z_+ .$$

Such a triple $(\mathcal{K}, J, U)$ is called a unitary power dilation of T. It follows that

$$p(T) = J^* p(U) J .$$

The inequality holds for all $p(U)$ since U is normal and thus for $p(T)$ since J and J^* are norm decreasing.

While every contraction possesses a unitary dilation, the proof is particularly easy in case $\|T\|_{op} < 1$.

Let $\mathcal{D}$ be the Hilbert space $\mathcal{H}$ but with the new inner product

$$(\xi,\eta)_{\mathcal{D}} = (\xi,\eta) - (T\xi, T\eta) .$$

For $\mathcal{K}$ we take $\ell^2(\mathbb{Z};\mathcal{D})$, define J by

$$(J\xi)(n) = T^n\xi \quad \text{if} \quad n \geq 0$$

$$= 0 \quad \text{if} \quad n < 0$$

and define U the backwards shift by

$$Uh(n) = h(n+1) .$$

A boring but easy calculation shows that

$$J^*h = \sum_{n=0}^{\infty} T^{*n}(I-T^*T)h(n)$$

and it follows that $T^n = J^*U^nJ$ for $n \in \mathbb{Z}_+$. In particular if $n = 0$, $I = J^*J$ so that J is an isometry.

Finally if $\|T\| \leq 1$ we can apply the foregoing to rT where $0 \leq r < 1$ to obtain

$$\|p(rT)\|_{op} \leq \|p\|_\infty .$$

Letting r tend to 1 gives the Von Neumann inequality.

§ 3. The stars left calculus

In cases where T is not normal, we clearly run into the problem that T and T^* do not commute. One way of keeping accounts so as to avoid this difficulty is to put all the T^*'s on the left and all the T's on the right. Of course some flexibility is lost in doing this and some results (Ando's inequality) cannot be obtained by these methods, but nonetheless it is a powerful tool.

We now return to the first step of Von Neumann's proof (which is an example of stars left calculus) and attempt to make it work for a general p. Let p be a disc algebra element with $\|p\|_\infty \leq 1$ and let $P = p(T)$. Then we wish to write something like

$$I - P^*P = \sum_k A_k^*(I-T^*T)A_k .$$

According to our philosophy A_k should be a polynomial in T (so that A_k^* is a polynomial in T^*). In this case we represent A_k by a Cauchy kernel

$$A_k = \int (I-\bar{\mu}T)^{-1} f_k(\mu)d\mu$$

(so that if f_k is an analytic polynomial, $A_k = f_k(T)$). In this way we have

$$(*) \qquad I - P^*P = \int (I-\lambda T^*)^{-1}(I-T^*T)(I-\bar{\mu}T)^{-1}K(\lambda,\mu)d\lambda d\mu$$

where $K(\lambda,\mu) = \sum_k \overline{f_k(\lambda)}f_k(\mu)$. ($d\lambda$, $d\mu$ represent Haar measure on the unit circle.) Clearly K should be positive definite, analytic in μ and conjugate analytic in λ. Indeed if K does have these properties the positive definiteness of $I - P^*P$ follows easily from that of $I - T^*T$. A calculation shows that for $(*)$ to hold we take

$$K(\lambda,\mu) = \frac{1-\overline{p(\lambda)}p(\mu)}{1-\bar{\lambda}\mu}$$

and it remains to show that this K is positive definite. Putting $p = 0$ in K leads to $(1-\bar{\lambda}\mu)^{-1}$ the defining kernel for H^2. With this in mind we see that

$$\int K(\lambda,\mu)f(\lambda)\overline{f(\mu)}d\lambda d\mu = \|f\|^2_{H^2} - \|J^*(\bar{p}f)\|_{H^2}$$

where J^* is the orthogonal projection of L^2 onto H^2. It is now easy to see that K positive definite is equivalent to $\|p\|_\infty \leq 1$.

The same game can be applied in countless more general situations. For instance let T_1 and T_2 commute, we wish to have Von Neumann's inequality hold with respect to the bidisc algebra norm. We shall need to use the Cauchy kernel $(1-\bar{\lambda}_1\mu_1)^{-1}(1-\bar{\lambda}_2\mu_2)^{-1}$ in place of $(1-\bar{\lambda}\mu)^{-1}$ and the hypothesis will have to be

$$I - T_1^*T_1 - T_2^*T_2 + T_1^*T_2^*T_1T_2 \quad \text{positive definite.}$$

(Ando's inequality shows that this result is not the best.) For J commuting operators $T_1 \ldots . T_J$ the criterion will be Δ positive definite where Δ is the expression obtained by rewriting $\prod_{j=1}^{J}(I-T_j^*T_j)$ "with all the stars of the left." In particular if $T_1 \ldots . T_J$ are commuting isometries, then Von Neumann's inequality holds. (In practice one needs to work with $rT_1 \ldots . rT_J$, $0 \leq r < 1$.)

We proceed now to a more complicated example [5]. Let us assume that $T_1 \ldots . T_J$ commute and that they satisfy the condition

$$\sum_{j=1}^{J} \|T_j\xi\|^2 \leq \|\xi\|^2$$

which is equivalent to the stars left constraint

$$I - T^*\circ T = I - \sum_{j=1}^{J} T_j^*T_j \quad \text{positive definite.}$$

Obviously now the unit ball in J dimensional complex space is relevant and the Cauchy (Cauchy-Szego) kernel for this is $(1-\bar{\lambda}\cdot\mu)^{-J}$ with respect to the normalized invariant measure σ on the sphere. We use the formula

$$I - P^*P = \int (I-\lambda\cdot T^*)^{-J}(I-T^*\cdot T)(I-\bar{\mu}\cdot T)^{-J}K(\lambda,\mu)\,d\sigma(\lambda)\,d\sigma(\mu)$$

where $K(\lambda,\mu) = (1-\bar{\lambda}\cdot\mu)^{-1}(1-\overline{p(\lambda)}p(\mu))$. We seek to understand what it means for K to be positive definite. Now $(1-\bar{\lambda}\cdot\mu)^{-1}$ is the defining kernel of a Hilbert space L (different from H^2, $J \geq 2$). There is a kernel $S(\lambda,\mu)$, analytic in μ, conjugate analytic in λ, the positive definite square root of $(1-\bar{\lambda}\cdot\mu)^{-1}$ with respect to H^2

$$(1-\bar{\lambda}\cdot\mu)^{-1} = \int \overline{S(\nu,\lambda)}S(\nu,\mu)\,d\sigma(\nu) \; .$$

The condition

$$\int K(\lambda,\mu)q(\lambda)\overline{q(\mu)}\,d\sigma(\lambda)\,d\sigma(\mu) \geq 0$$

now amounts to

$$\| \int S(\cdot,\mu)p(\mu)\overline{q(\mu)}\,d\sigma(\mu)\|_2 \leq \| \int S(\cdot,\mu)\overline{q(\mu)}\,d\sigma(\mu)\|_2 \, .$$

Now define a Hilbert space K to consist of functions f which admit a representation

$$f(\mu) = \int S(\nu,\mu)\phi(\nu)\,d\sigma(\nu) \quad \text{for} \quad \phi \in H^2 .$$

Thus K is just the dual of L viewed through the H^2 inner product. Our condition states by duality that

$$\sup_{\|f\|_K \leq 1} \left|\int f(\mu)p(\mu)\overline{q(\mu)}\,d\sigma(\mu)\right| \leq \sup_{\|f\|_K \leq 1} \left|\int g(\mu)\overline{q(\mu)}\,d\sigma(\mu)\right|$$

for every q. Let us normalize q so that $\|q\|_L = 1$ then the R.H.S. is just 1. The L.H.S. is then clearly the norm of p as a multiplier of K. Thus letting

$$\|p\|_{\text{mult}(K)} = \sup_{\|f\|_K \leq 1} \|pf\|_K$$

we have

$$\|p(T_1\cdots T_J)\| \leq \|p\|_{\text{mult}(K)} \, .$$

This result is best possible in the following sense let $Z_1 \cdots Z_J$

denote multiplication by $z_1 \ldots z_J$ on K (resp). Then $z_1^* \ldots z_J^*$ satisfy the condition on $T_1 \ldots T_J$. Also it is not too hard to see that

$$\|p(z_1^* \ldots z_J^*)\| = \|p\|_{\mathrm{mult}(K)} .$$

Just for the record K consists of those functions f whose power series expansion

$$f = \sum_{n \in \mathbb{Z}_+^J} a_n z^n$$

satisfies $\|f\|_K^2 = \sum_{n \in \mathbb{Z}_+^J} |a_n|^2 (\beta(n))^{-1} < \infty$

where $\beta(n) = (n_1 + \ldots + n_J)! \{n_1! \ldots n_J!\}^{-1}$ are the multinomial coefficients.

4. Several commuting contractions

Let $T_1, \ldots, T_J$ be commuting contractions. Then in general the inequality

$$\|p(T_1 \ldots T_J)\| \leq \|p\|_\infty$$

fails for $J \geq 3$ (Here $\|p\|_\infty$ is the sup over the polydisc.) This is seen most easily with the example of Crabb and Davie [6]. For the essentially typical case $J = 3$, H is the 8-dimensional space with orthonormal basis $e, f_1, f_2, f_3, g_1, g_2, g_3, h$. Define

$$T_i e = f_i, \quad T_i f_i = -g_i, \quad T_i f_j = g_k \quad (i,j,k \text{ all different}),$$

$$T_i g_j = \delta_{ij} h, \quad T_i h = 0.$$

It is easy to see that T_1, T_2, T_3 are indeed commuting contractions and that $T_1^3 = T_2^3 = T_3^3 = -T_1 T_2 T_3$ is the operator of norm one which takes e to $-h$ and every other element of the basis to zero. Let

$$p(z_1,z_2.z_3) = z_1^3 + z_2^3 + z_3^3 - 3z_1z_2z_3 .$$

Then $\|p(T_1,T_2,T_3)\|_{op} = 6$. Clearly $\|p\|_\infty \leq 6$ and if $\|p\|_\infty = 6$ then there exist z_1,z_2,z_3 of modulus 1 such that $z_1^3 = z_2^3 = z_3^3 = -z_1z_2z_3$. Thus $(-z_1z_2z_3)^3 = z_1^3z_2^3z_3^3$ so that $-1 = 1$. Hence $\|p\|_\infty < 6$.

It is not known what is the smallest dimension in which (**) fails. We can show that it holds in dimension 2. Let $T_1 \ldots T_J$ be commuting contractions on a 2-dimensional Hilbert space. A calculation with 2×2 matrices shows that the commutant of a matrix A consists of matrices of the form $\alpha I + \beta A$ $(\alpha,\beta \in C)$ unless A is a scalar multiple of I (in which case the commutant consists of all 2×2 matrices). One deduces that either $T_1,\ldots,T_J$ are all scalar multiples of the identity (in which case our inequality is trivial) or there is an endomorphism T scalars α_j, β_j such that $T_j = \alpha_j I + \beta_j T$. Now find a sequence $T^{(k)}$ of semisimple (diagonalizable) endomorphisms converging to T and define

$$T_j^{(k)} = \frac{\|\alpha_j I + \beta_j T\|}{\|\alpha_j I + \beta_j T^{(k)}\|} (\alpha_j I + \beta_j T^{(k)}) .$$

Then for fixed k, $T_1^{(k)} \ldots T_J^{(k)}$ is a set of commuting contractions and $T_j^{(k)} \to T_j$ as $k \to \infty$ with j fixed. In consequence it suffices to prove the inequality for semisimple commuting contractions and we assume henceforth that $T_1 \ldots T_J$ have this property. Hence there is a basis e_1,e_2 normalized $(\|e_1\| = \|e_2\| = 1)$ (e_1,e_2) real ≥ 0 but not in general orthogonal with respect to which T_j is represented by the matrix $\begin{pmatrix}\lambda_j & 0 \\ 0 & \mu_j\end{pmatrix}$. Let p be the polydisc algebra element with $\|p\|_\infty \leq 1$. By applying Möbius transformations both to each variable of the domain of p and to the range of p we may make the following assumptions.

$$\mu_j = 0 \quad (1 \le j \le J) \quad p(0,\ldots,0) = 0 .$$

Now let $(e_1,e_2) = \cos\theta$. A calculation shows that T_j is a contraction is equivalent to

$$|\lambda_j| \le \sin\theta \qquad (0 < \theta \le \pi/2) .$$

Of course $P = p(T_1 \ldots T_J)$ has matrix

$$\begin{pmatrix} \nu & 0 \\ 0 & 0 \end{pmatrix}$$

where $\nu = p(\lambda_1 \ldots \lambda_J)$. We wish to show that P is a contraction that is $|\nu| \le \sin\theta$ and this is an immediate consequence of the Schwartz inequality.

Another situation in which (**) holds is the case $J = 2$. That is known as Ando's inequality [7]. Let us give the proof on the basis that H is finite dimensional, and indicate later the reduction to this case. Thus we assume that T_1, T_2 are commuting contractions on a finite dimensional Hilbert space H. Let $D_j = (I-T_j^*T_j)^{1/2}$. We define V_j on $\ell^2(\mathbb{Z}_+,H)$ by

$$V_j(\xi_0,\xi_1,\xi_2,\ldots) = (T_j\xi_0,\ D_j\xi_0,\ \xi_1,\ \xi_2,\ldots) .$$

It is easily verified that V_j is an isometry. In fact V_j is an isometric dilation of T_j. We set

$$J\xi = (\xi,0,\ldots\ldots) .$$

One easily checks that $J^*V_j^nJ = T_j^n \quad \forall n \in \mathbb{Z}_+$. Unfortunately V_1 and V_2 do not commute and so no conclusions can be drawn as yet.

$$V_1V_2(\xi_0,\xi_1,\xi_2,\ldots) = (T_1T_2\xi_0,D_1T_2\xi_0,D_2\xi_0,\xi_1,\xi_2,\ldots)$$

$$V_2V_1(\xi_0,\xi_1,\ldots) = (T_2T_1\xi_0,D_2T_1\xi_0,D_1\xi_0,\xi_1,\xi_2,\ldots) .$$

A calculation shows that

$$\|D_1T_2\xi_0\|^2 = \|D_2\xi_0\|^2 = \|D_2T_1\xi_0\|^2 + \|D_1\xi_0\|^2 .$$

Let us define maps $A_j : H \to H\oplus H$ by

$$A_1\xi = (D_1T_2\xi, D_2\xi), \qquad A_2\xi = (D_2T_1\xi, D_1\xi) .$$

We know that there is an isometry Γ

$$\Gamma: \quad im(A_1) \to im(A_2)$$

s.t.

$$\Gamma A_1 = A_2 .$$

Also we know that since $\ker(A_1) = \ker(A_2)$ and in view of the finite dimensionality, $\dim im(A_1) = \dim im(A_2)$. Hence Γ can be extended to a unitary (that we will still denote Γ on the whole of $H\oplus H$.

Now define

$$G(\xi_0,\xi_1,\xi_2,\ldots\ldots) = (\xi_0,\Gamma(\xi_1,\xi_2),\Gamma(\xi_3,\xi_4),\ldots.)$$

so that G is unitary on $\ell^2(\mathbb{Z}_+,H)$.

Put $U_1 = GV_1$, $U_2 = V_2G^{-1}$: calculations show that U_1 and U_2 commute are isometries and are still a power dilation in the sense

$$T^n = J^*U^nJ \qquad n \in \mathbb{Z}_+^2 .$$

The inequality now follows from that for commuting isometries.

§ 5. Reduction to finitely many dimensions

For p a polynomial in J variables define

$$\|p\|_{VN} = \sup_A \|p(T_1\ldots.T_J)\|_{op}$$

$$\|p\|_{FVN} = \sup_{\mathcal{B}} \|p(T_1 \ldots . T_J)\|_{op}$$

$$\|p\|_{NVN} = \sup_{\mathcal{C}} \|p(T_1 \ldots . T_J)\|_{op}$$

where $\mathcal{A}(\mathcal{B})$ is the set of all J-tuples of commuting contractions $(T_1 \ldots . T_J)$ on (finite dimensional) Hilbert space and $\mathcal{C}$ is the subset of $\mathcal{B}$ consisting of nilpotent J-tuples. It is our object to show that these norms are all equal. An immediate consequence of this is the semisimplicity of the algebra VN_J (that is the completion of polynomials in $\| \ \|_{VN}$). This was observed earlier by Varopoulos using completely different ideas, [8].

The difficulty in reducing is to find operators on a finite dimensional space which still commute. We proceed as follows. Let $X_M = \{m; m \in \mathbb{Z}, |m| \leq M\}$ and denote by S the forwards shift on $\ell^2(X_M)$. Then $A_j = S \times T_j$ are commuting contractions on $\ell^2(X_M, \mathcal{H})$. Let $\xi \in \mathcal{H}$ with $\|\xi\|_{\mathcal{H}} < 1$, p a polynomial

$$p(z) = \sum_{|n| \leq N} p_n z^n$$

with $\|p\|_{NVN} \leq 1$ and let $f = (2M+1)^{-1/2} \mathbb{1}_{X_M}$. Let $\mathcal{F} = sp\{A^n(f \otimes \xi)\}$ a finite dimensional subspace of $\ell^2(X_M, \mathcal{H})$ by virtue of the nilpotency of S and of A_j. Clearly $\mathcal{F}$ is invariant under A and we may regard the A_j as operators on $\mathcal{F}$. Hence

$$\|p(A)(f \otimes \xi)\| \leq 1 .$$

Now let η be arbitrary in $\mathcal{H}$ with $\|\eta\|_{\mathcal{H}} \leq 1$. Then

$$|(p(A) f \otimes \xi, f \otimes \eta)| \leq 1$$

and a calculation shows that this gives

$$|(K_M p(T) \xi, \eta)| \leq 1$$

where $K_M p$ is the polynomial given by

$$K_M p(z) : \sum_{|n| \leq N} (1 - \frac{|n|}{2M+1}) p_n z^n$$

at least for $M > N$. Taking sups over Γ, ξ and η we get $||K_M p||_{VN} \leq 1$. Clearly $||q||_{VN} \leq ||q||_A$ and so we find

$$||p||_{VN} \leq 1 + ||p - K_M p||_{VN} \leq 1 + ||p - K_M p||_A .$$

Now let $M \to \infty$ and we find $||p||_{VN} \leq 1$. The other inequalities

$$||p||_{NVN} \leq ||p||_{FVN} \leq ||p||_{VN}$$

are evident.

§ 6. A new proof of Ando's inequality for homogenous polynomials

Using the ideas of the previous section we see that it is efficient to prove our result for graded contractions. That is

$$\mathcal{H} = \bigoplus_{n \in \mathbb{Z}} \mathcal{H}_n \qquad \text{(orthogonal direct sum)}$$

T_1, T_2 are commuting contractions and T_j maps $\mathcal{H}_n$ into $\mathcal{H}_{n+1}$. Since p is homogeneous say of degree k, $p(T_1,T_2)$ maps $\mathcal{H}_n$ into $\mathcal{H}_{n+k}$ and we need only prove that it is a contraction on each level separately. Thus without loss of generality we need only consider the case $m = 0$. Thus we take a unit vector e_0 (the subscript is the zero multi-index) in $\mathcal{H}_0$ and show that $||p(T_1,T_2)e_0|| \leq 1$. Let now $e_a = T^a e_0 \in \mathcal{H}_{|a|}$. Again without loss of generality we can assume that $\mathcal{H}_m$ is spanned by e_a for $|a| = m$. Define now $q_m(a,b) = (e_a, e_b)$ for $|a| = |b| = m$. Then the matrix a_m gives the geometry on $\mathcal{H}_m$. The fact that T_j is a contraction amounts to the positive definiteness of the matrices

$$q_m(a,b) - q_{m+1}(a+\mathbf{1}_j,\ b+\mathbf{1}_j)\ .$$

The idea is to define new matrices r_m such that $r_0 = q_0$, $r_m \geq q_m$ (in the p.d. sense) in such a way that in the new geometry T_j is an isometry. Then we have

$$\|p(T_1,T_2)e_0\|_{old} \leq \|p(T_1,T_2)e_0\|_{new} \leq \|e_0\|_{new} = \|e_0\|_{old}\ .$$

The L.H. inequality holds since $q_k \leq r_k$, the R.H. since $r_0 = q_0$ and the middle since T_j are isometries. This procedure is essentially a variant of the dilation idea. We will construct the r's inductively by means of the following lemma. We state the lemma in greater generality than we need.

Lemma. Let V be a finite dimensional vector space over C, W a subspace of codimension J. Let $f_1 \ldots . f_J$ be a basis of V modulo W and let $V_j = sp(W,f_j)$. Let α_j be a positive definite (in the wide sense) sesquilinear form on V_j such that α_j and α_k agree on W. Then there exists a positive definite sesquilinear form α on V such that α and α_j agree on V_j.

Proof. We give here the proof only in the case that each α_j is strictly positive and leave the exceptional cases to the reader. Choose a basis in W so that the commonly defined form is represented by the identity matrix and renormalize the f_j so that $\alpha_j(f_j,f_j) = 1$. Then

$$[\alpha_j] = \left(\begin{array}{c|c} 1 & \bar{\xi}_1^{(j)} \ldots\ldots\ldots\ldots \bar{\xi}_K^{(j)} \\ \hline \xi_1^{(j)} & \\ \vdots & I \\ \xi_K^{(j)} & \end{array}\right) \begin{array}{l} \}\ f_j \\ \\ \Big\}\ W \\ \\ \end{array}$$

for certain scalars $\xi_k^{(j)}$ $1 \leq j \leq J$, $1 \leq k \leq \dim(W) = K$. Positive definiteness of $[\alpha_j]$ is then equivalent to $\sum_{k=1}^{K} |\xi_k^{(j)}|^2 \leq 1$. We define α by

$$[\alpha] = \left(\begin{array}{cccc|c} 1 & \xi^{(1)*}\xi^{(2)} & \cdots & \xi^{(1)*}\xi^{(J)} & \xi^{(1)*} \\ \xi^{(2)*}\xi^{(1)} & 1 & & & \xi^{(2)*} \\ \cdots & & & & \cdots \\ \xi^{(J)*}\xi^{(1)} & & & 1 & \xi^{(J)*} \\ \hline \xi^{(1)} & \xi^{(2)} & \cdots & \xi^{(J)} & I \end{array}\right)$$

That is the "missing" entries $\alpha(f_j, f_k)$ are given by $\xi^{(j)*}.\xi^{(k)}$. The positive definiteness of α comes from the formula $[\alpha] = P^*\Delta P$ where

$$\Delta = \left(\begin{array}{ccc|c} 1-\|\xi^{(1)}\|^2 & & 0 & \\ & 1-\|\xi^{(2)}\|^2 & & 0 \\ 0 & & 1-\|\xi^{(J)}\|^2 & \\ \hline & 0 & & I \end{array}\right)$$

$$P = \left(\begin{array}{cccc|c} & & 0 & & 0 \\ \hline \xi^{(1)} & \xi^{(2)} & \cdots & \xi^{(J)} & I \end{array}\right).$$

Now let us define the r_m inductively. Let $r_0 = q_0$. This starts the induction. Now suppose that r_m has been defined with $r_m \geq q_m$ and so that $r_{m-1}(c,d) = r_m(c + \mathbb{1}_j,\ d + \mathbb{1}_j)$. We come to define r_{m+1}. We shall have to set

$$r_{m+1}(a + \mathbb{1}_j,\ b + \mathbb{1}_j) = r_m(a,b)$$

and this is well defined by the inductive hypothesis. Since $r_m \geq q_m$ and since

$$q_m(a,b) - q_{m+1}(a + \mathbb{1}_j,\ b + \mathbb{1}_j)$$

is positive definite

$$r_{m+1}(a+\mathbb{1}_j,\ b+\mathbb{1}_j) - q_{m+1}(a+\mathbb{1}_j,\ b+\mathbb{1}_j)$$

is positive definite. Since the codimension condition is satisfied one now uses the lemma to extend $r_{m+1} - q_{m+1}$ to all pairs of multi-indices of length $m+1$ in such a way that $r_{m+1} - q_{m+1}$ is positive definite. This completes the proof. The argument does not generalize to $J \geq 3$ as the codimension condition is violated.

§ 7. The Kaijser-Varopoulos counterexample and scrambled shifts

The Kaijser-Varopoulos counterexample [9] achieves the same end as the Crabb and Davie counterexample [6]. Here H is a 5-dimensional Hilbert space with orthogonal basis e, e_1, e_2, e_3, f. The operators T_j $(j = 1,2,3)$ are defined by $T_i e = e_i$, $T_i e_j = a_{ij} f$. For a we take

$$a = 3^{-\frac{1}{2}} \begin{pmatrix} 1 & -1 & -1 \\ -1 & 1 & -1 \\ -1 & -1 & 1 \end{pmatrix}.$$

It is easy to see that T_j are commuting contractions. The polynomial is defined by the same matrix

$$p(z) = \sum_{j,k} \overline{a_{jk}}\ z_j\ z_k$$

so that $(p(T)e,f) = \sum_{j,k} |a_{jk}|^2 = 3.$

A calculation shows that $\|p\|_\infty < 3.$

We set out now to show that the obvious generalization of this example satisfies

$$\|p(T_1 \ldots T_J)\| \leq C_J \|p\|_\infty$$

where C_J is a constant depending only on the number J of contractions.

Let e_n, $|n| \leq K$ and f_n, $|n| \leq K-1$ form an orthonormal basis. We group elements to form a grading. Thus

$$E_k = \mathrm{sp}\{e_n;\ |n| = k\} \qquad 0 \leq k \leq K$$

$$F_k = \mathrm{sp}\{f_n;\ |n| = k\} \qquad 0 \leq k \leq K-1\ .$$

For $0 \leq k \leq K-1$, T_j is a forward shift on E_k

$$T_j e_n = e_{n+\mathbb{1}_j} \qquad |n| \leq K-1\ .$$

For $0 \leq k \leq K-1$, T_j is a backwards shift on F_k

$$\begin{aligned} T_j f_n &= f_{n-\mathbb{1}_j} && \text{if} \quad n_j > 0 \\ &= 0 && \text{if} \quad n_j = 0\ . \end{aligned}$$

As an intermediary consider F_K and define S the scrambling operator $S\colon E_K \to F_K$. Let it have matrix $s(p,q)$ so that

$$Se_q = \sum s(p,q) f_p\ .$$

Then on E_K we define $T_j\colon E_K \to F_{K-1}$ by $T_j = V_j S$ where V_j is the backwards shift $F_K \to F_{K-1}$.

$$T_j e_q = \sum s(p+\mathbb{1}_j,\ q) f_p\ .$$

Let us denote $s_j(p,q) = s(p+1_j,\ q)$, $p \in I_{K-1}$, $q \in I_K$. Clearly T_j is a contraction iff s_j is and the commutativity amounts to the fact that $s(p,q)$ is a function of $p+q$

$$s(p,q) = \sigma(p+q)\ .$$

Here σ is defined on I_{2K}.

<u>Lemma</u>. $||s||_{op} \leq J^{1/2}$.

<u>Proof</u>. Let $b \in \ell^2(I_K)$ and let $a(p) = \sum_q s(p,q)b(q)$. Then since $||s_j||_{op} \leq 1$, $||a_j|| \leq ||b||$ where $a_j(p) = a(p+\mathbb{1}_j)$. Then

$$\sum_{p\in I_K} |a(p)|^2 \leq \sum_{j=1}^{J} \sum_{p\in I_{K-1}} |a_j(p)|^2 \leq J||b||^2$$

and the result follows.

By splitting the space up into pieces we see that we need only prove

$$|(\phi(T_1 \ldots T_J)\xi,\eta)| \leq C_J ||\phi||_\infty ||\xi|| \; ||\eta||$$

where $\xi \in \bigoplus_{k=0}^{K} E_k$ & $\eta \in \bigoplus_{k=0}^{K-1} F_k$. In fact let

$$\phi(z) = \sum \phi_n z^n, \; \xi = \sum \xi_a e_a, \quad \eta = \sum \eta_b f_b .$$

Then

$$(\phi(T_1 \ldots T_J)\xi,\eta) = \sum_{n,a,b} \xi_a \bar{\eta}_b \phi_n \sigma(a+b+n) .$$

We now define $\psi(z_1 \ldots z_J) = \phi(z_1 \ldots z_J)(\Sigma\xi_a z^a)$.

Then $||\psi||_2 \leq ||\phi||_\infty ||\xi||$ and if we denote

$$\psi(z) = \sum \psi_m z^m .$$

We have $(\phi(T_1 \ldots T_J)\xi,\eta) = \sum_{b,m} \psi_m \bar{\eta}_b \sigma(b+m)$.

In principle ψ contains all homogeneities, but we throw away all except those in the range

$$K + 1 \leq |m| \leq 2K$$

since these are the only ones which contribute to the inner product.

This can only reduce $\|\psi\|_2$. Also we may consider just one homogeneity at a time. Our problem becomes as follows. For each k, $0 \le k \le K-1$, $\eta \in F_k$ and $\psi \in H^2$ homogeneous of degree $2K-k$ we must prove

$$\left| \sum_{b,m} \psi_m \bar{\eta}_b \, \sigma(b+m) \right| \le C_J \|\psi\| \; \|\eta\|$$

<u>Lemma</u>. $N = N_J$ an integer s.t. for all k, $0 \le k \le K-1$, a set $\{a_1 . . . . . a_N\} - I_{K-k}$ s.t.

$$\sum_{n=1}^{N} a_n + I_K = I_{2K-k} \; .$$

<u>Proof</u>. Exercise.

<u>Corollary</u>. We may write

$$\psi(z) = \sum_{n=1}^{N_J} m_n(z) \psi_n(z)$$

where m_n is a monomial of degree $K-k$ and ψ_n is homogenous of degree K. Further

$$\sum \|\psi_n\|_2 \le N_J^{1/2} \|\psi\|_2 \; .$$

Splitting up the inner product according to this decomposition and using the estimate on $\|s\|_{op}$ we have the required result.

I should like to take this opportunity to thank the University of Connecticut for its hospitality during the special year in Harmonic Analysis.

References

[1] J. Von Neumann, Eine Spectraltheorie für allgemeine Operatoren eines unitären Raumes, Math Nachr 4 (1951) 258-281.

[2] S. Fisher, The convex hull of finite Blashke products, Bull A.M.S. 74 (1968) 1128-1129.

[3] B. Sz-Nagy and C. Foias, Harmonic Analysis of Operators on Hilbert Space, North-Holland 1970.

[4] P. R. Halmos, Shifts on Hilbert Space, J. reine angew Math 208 (1961) 102-112.

[5] S. W. Drury, A Generalization of Von Neumann's inequality to the complex ball, Proc A.M.S., 68/3 (1978) 300-304.

[6] M. J. Crabb and A. M. Davie, Von Neumann's inequality for Hilbert Space operators, Bull London Math Soc. 7 (1975) 49-50.

[7] T. Ando, On a pair of commutative contractions, Acta Sci. Math 24 (1963) 88-90.

[8] N. Th. Varopoulos, On a class of Banach Algebras, Lecture Notes in Math. No. 512, (1975) 180-184.

[9] N. Th. Varopoulos, On an inequality of Von Neumann and an application of the metric theory of tensor products to operator theory, J. Func. Analysis, 16 (1974) 83-100.

A SIMPLE-MINDED PROOF OF THE PISIER-GROTHENDIECK INEQUALITY

Sten Kaijser

0. Introduction

The purpose of this paper is to give a new proof of Pisier's theorem for bilinear forms on a C*-algebra. Since a C(X)-space is a commutative C*-algebra, Pisier's theorem contains Grothendieck's "fundamental theorem" on tensor products. Our new proof of Pisier's theorem is therefore also a proof of Grothendieck's theorem and we believe that it is new also as a proof of this theorem. If so, it seems to be a very simple (not to say simple-minded) proof of that theorem. It also seems to be quite illuminating in the sense that it explains why other proofs work. In one way or another all proofs of Grothendieck's theorem are based on the fact that if E is a subspace of $L^q(\Omega,P)$ ($q > 2$, P a probability measure) on which the L^q-norm and the L^2-norm are equivalent, then every $f \in E$, can be decomposed as

$$f = g + h \tag{0.1.1}$$

where

$$g \in L^\infty \quad \text{and} \quad h \in L^2 ,$$

in such a way that g is the 'significant part' of f in the sense that the L^2-norm of h is small. In the original proof h is simply forgotten; in the proofs based on the disc algebra one exploits the possibility of choosing g either analytic or antianalytic and in the proofs based on interpolation this decomposition is only implicit.

(For the original proof see [3], [5], [6], for the proofs using the disc algebra see [2] and [9] and for the proofs using interpolation see [4], [8] and [10].)

However, as soon as one realizes that the decomposition (0.1.1) is the common feature of all these proofs, then the simplest way to obtain the decomposition is of course to simply truncate the function f at a suitable height and this is what we do in our proof. Now the proofs based on interpolation really use the fact that this decomposition can be done at all heights and using this additional information it is clear that the proofs based on interpolation should give better estimates of the constants. It is therefore somewhat surprising that in the non-commutative case our proof gives an estimate of the constant which is slightly better than Pisier's estimate. Pisier gets the bound $K_P \leq 12$ while our method gives the bound $K_P < \frac{81}{8}$.

Finally, a comparison of our proof and Pisier's proof yields the following similarities and differences. Formally Pisier studies operators from a C*-algebra into a Banach space of cotype 2 while we study bilinear forms on a pair of C*-algebras. At least formally Pisier therefore proves a stronger theorem than we do even if the difference is basically a formality. The main similarities are that we shall use the same facts about C*-algebras as Pisier does - including the estimates for the Rademacher functions. The main difference between the proofs is that our proof is based on C*-algebra techniques (a key step is the use of spectral theory through the functional calculus of Hermitian elements) while we do not need the methods of Banach space theory.

To conclude this introduction we finally want to remark that our proof (as well as Pisier's) only involves the self-adjoint parts of the given C*-algebras, and since Pisier very clearly explains the passage to the entire algebras, we shall not even state the theorem in that case.

1. Notations and preliminaries

1. Let $A_h(B_h)$ be the set of Hermitian elements in the C*-algebra A(B). We shall write $C = A_h \otimes B_h$ to denote their algebraic tensor product. We shall use the following definitions.

Definition (1.1). Let V be the set of all $v \in \mathbb{C}$ such that there exist finite sets $\{a_i\} \subset A_h$, $\{b_i\} \subset \mathbb{B}_h$ so that

$$v = \Sigma\, a_i \otimes b_i$$

and

$$\Sigma\, \|a_i\| \cdot \|b_i\| \leq 1 .$$

Definition (1.2). Let U be the set of all $u \in \mathbb{C}$ such that there exist finite sets $\{a_i\} \subset A_h$, $\{b_i\} \subset \mathbb{B}_h$ so that

$$u = \Sigma\, a_i \otimes b_i$$

and

$$\|\Sigma\, a_i^2\| \leq 1 , \quad \|\Sigma\, b_i^2\| \leq 1 .$$

It is clear that both V and U are convex symmetric subsets of $\mathbb{C}$ and that $V \subset U$. We shall soon see that the set U is bounded and since the set V is absorbing the Minkowski-functionals p_V and p_U (i.e. $p_V(w) = \inf\{t \mid w \in tV\}$) are both norms on $\mathbb{C}$. The norm p_V is of course simply the projective tensor norm on $\mathbb{C}$ while the norm p_U is a reasonable norm on $\mathbb{C}$.

Remark. It follows from a conjecture of Ringrose, which, as Pisier proved, follows from his theorem, that the norm p_U is in fact equivalent to a tensor norm on $\mathbb{C}$.

It is a well-known consequence of the Gelfand-Naimark-Segal theorem that every C*-algebra has an isometric representation as a closed subalgebra of the space $B(H)$ of all bounded operators on some Hilbert space H (see e.g. [7]). Since we shall not perform any operations that will lead us outside the C*-algebra generated by a given set of elements we shall, in the following, assume that $A = B(H)$ while $\mathbb{B} = B(K)$. It follows that $A_h \otimes \mathbb{B}_h \subset B(H \otimes K)_h$ where $B(H \otimes K)_h$

is the set of Hermitian elements in the algebra of all bounded operators on the Hilbert space $H \otimes K$ (denoting the Hilbert space tensor product). The fact that U is a bounded subset of $\mathbb{C}$ follows now from the following.

<u>Proposition (1.3)</u>. Let U be as in Definition (1.2) and let $u \in U$. Then

$$\|u\|_{B(H \otimes K)} \leq 1 .$$

<u>Proof</u>. Since u is Hermitian it suffices to prove that

$$|(ux|x)| \leq \|x\|^2$$

for all $x \in H \otimes K$ and it follows from simple properties of the Hilbert space tensor product that it suffices to prove this for the simple tensors $x = h \otimes k$. In that case we have however

$$\big(u(h \otimes k)|(h \otimes k)\big) = \big(\Sigma(a_i \otimes b_i)(h \otimes k)|(h \otimes k)\big) =$$

$$= \Sigma\ (a_i h|h)(b_i k|k) \leq (\Sigma\ \|a_i h\| \cdot \|b_i k\|) \cdot \|h\| \cdot \|k\|$$

and assuming $\|h\| = \|k\| = 1$ we then have

$$(\Sigma\ \|a_i h\| \cdot \|b_i k\|) \leq (\Sigma\ \|a_i h\|^2)^{1/2} \cdot (\Sigma\ \|b_i k\|^2)^{1/2}$$

and it suffices now to observe that

$$\Sigma\ \|a_i h\|^2 = \Sigma(a_i^2 h|h) = \big((\Sigma\ a_i^2)h|h\big) \leq 1 .$$

2. Let $\mathbb{C}_V(\mathbb{C}_U)$ be the space $\mathbb{C}$ endowed with the norm p_V (p_U). The dual space $\mathbb{C}'_V$ is then the space of all bounded bilinear forms on $A_h \times B_h$ while the elements of the dual space $\mathbb{C}'_U$ will be called $2-C^*$-summing bilinear forms.

Remark. It is easily seen that this is a special case of Pisier's notion of $2-C^*$-summing operators. Pisier also proved that if T is an operator from a C^*-algebra having $2-C^*$-summing norm 1, then there exists a state of $f \in A'$, such that $\|Tx\|^2 \leq f(x^2)$, $x \in A_h$. It follows that if $\beta \in \mathbb{C}_U'$, $\|\beta\| \leq 1$, then there exist states $f \in A'$, $g \in \mathbb{B}'$, such that

$$|\beta(a,b)|^2 \leq f(a^2)\cdot g(b^2), \quad a \in A_h, \ b \in \mathbb{B}_h .$$

The main result proved in this paper is the following.

Theorem 1 (Pisier). Let $\beta \in \mathbb{C}_U'$, $\|\beta\|_{\mathbb{C}_U'} = 1$. Then

$$\|\beta\|_{\mathbb{C}_V'} \geq \frac{16}{81} .$$

Corollary. Let $\overline{\mathbb{C}}_V$ and $\overline{\mathbb{C}}_U$ denote the completions of the normed space $\mathbb{C}_V$ and $\mathbb{C}_U$. The canonical map from $\overline{\mathbb{C}}_V$ to $\overline{\mathbb{C}}_U$ (given by the extension of the identity map on $\mathbb{C}$) is then surjective and for every $x \in \overline{\mathbb{C}}_U$ there exists $y \in \overline{\mathbb{C}}_V$, such that

$$\|y\|_{\overline{\mathbb{C}}_V} \leq \frac{81}{16}\cdot \|x\|_{\overline{\mathbb{C}}_U} .$$

To prove Theorem 1 we shall use the following.

Main Lemma. Let $u \in U$. There then exist $v, w \in \mathbb{C}$, such that

$$u = v + w$$

and such that

$$p_V(v) \leq \frac{27}{16}, \quad p_U(w) \leq \frac{2}{3} .$$

(i.e. $U \subset \frac{27}{16} V + \frac{2}{3} U$).

Assuming the main lemma we shall first give the almost trivial

Proof of Theorem 1. Let $\varepsilon > 0$ be given. We are given $\beta \in \mathbb{C}'_U$, $\|\beta\|_{\mathbb{C}'_U} = 1$. There exists therefore $u \in U$, such that

$$\beta(u) > 1 - \varepsilon .$$

By the main lemma we then have

$$\beta(u) = \beta(v) + \beta(w)$$

and hence

$$|\beta(v)| > \frac{1}{3} - \varepsilon .$$

Since $p_V(v) \leq \frac{27}{16}$ it follows that $\|\beta\|_{\mathbb{C}'_V} \geq \frac{16}{81} - \varepsilon$. Since $\varepsilon > 0$ is arbitrary the theorem follows.

2. Proof of the main lemma

1. To prove our main lemma we shall need some auxiliary lemmas that will just be called lemmas in the following. First we shall introduce some notations.

For notational convenience we first prove the main lemma under the additional assumption that $A = B(H)$ while $B = B(K)$. By assumption we are given finite subsets $\{a_i\}_{i=1}^N \subset A_h$ and $\{b_i\}_{i=1}^N \subset \mathbf{B}$, such that

$$\Sigma\, a_i^2 \leq 1_H \quad (1_H \text{ being the identity operator on } H)$$

and

$$\Sigma\, b_i^2 \leq 1_K ,$$

and such that

$$u = \Sigma\, a_i \otimes b_i .$$

For notational convenience we shall also use Rademacher functions in the proof. Observe however that since we only have finite sets of elements, all integrals in the following are in fact simply finite sums. We now define

$$A = A(t) = \Sigma\, a_i r_i(t) \quad \big(B = B(t) = \Sigma\, b_k r_k(t)\big)\,.$$

In the following we shall consider A both as an operator on the Hilbert space $\mathbb{H} = L^2(\Pi, H)$ (Π denoting the unit interval) and as an element of $L^2(\Pi, A_h)$. Likewise B may be considered as belonging to either $B(\mathbb{K})$ or $L^2(\Pi, \mathbb{B}_h)$. We shall need the following.

<u>Lemma (2.1)</u>. Let $A = A(t)$ $\big(B = B(t)\big)$ be as above. Then

(i) $A(B)$ is a Hermitian element of $B(\mathbb{H})$ $\big(B(\mathbb{K})\big)$ (2.1.1)

(ii) $\int A(t) \otimes B(t)dt = \Sigma\, a_i \otimes b_i = u$ (2.1.2)

(iii) $\int A(t)^2 dt = \Sigma\, a_i^2 \le 1_H$ $\big(\int B(t)^2 dt \le 1_K\big)$ (2.1.3)

<u>Proof</u>. (i) is obvious, while (ii) and (iii) are formal identities using only the orthogonality (and not the independence) of the Rademacher functions.

A much more interesting result is the following.

<u>Lemma (2.2)</u> (Pisier). With the same notations as in the preceding lemma we also have

$$\int A(t)^4 dt \le 3\cdot 1_H \quad \big(\int B(t)^4 dt \le 3\cdot 1_K\big)\,. \tag{2.1.4}$$

As announced in the introduction our proof is based on truncation so we start by introducing the function a_τ defined by

$$a_\tau(\lambda) = \begin{cases} -\tau & \text{if } \lambda \le -\tau \\ \lambda & \text{if } -\tau \le \lambda \le \tau \\ \tau & \text{if } \tau \le \lambda \end{cases}$$

Since A (B) is a self-adjoint operator it has a spectral decomposition

$$A = \int \lambda dE(\lambda) \qquad \left(B = \int \mu dF(\mu)\right)$$

so we can define

$$A_\tau = \int a_\tau(\lambda) dE(\lambda) \qquad \left(B_\tau = \int a_\tau(\mu) dF(\mu)\right),$$

and $A^\tau = A - A_\tau \quad (B^\tau = B - B_\tau)$.

To complete the proof we shall now need one more lemma.

Lemma (2.3). With A_τ and A^τ (B_τ and B^τ) as above we have

(i) $A_\tau^2 \leq A^2$ and hence $\int A_\tau(t)^2 dt \leq 1_H$ (2.1.5)

$$\left(\int B_\tau(t)^2 dt \leq 1_K\right)$$

(ii) $(A^\tau)^2 \leq \frac{A^4}{16\tau^2}$ and hence $\int A^\tau(t)^2 dt \leq \frac{3}{16\tau^2} \cdot 1_H$ (2.1.6)

$$\left(\int B^\tau(t)^2 dt \leq \frac{3}{16\tau^2} \cdot 1_K\right)$$

Proof. (i) is obvious from the construction while (ii) follows from the following elementary inequality:

$$|t - a_\tau(t)| \leq \frac{t^2}{4\tau}. \tag{2.1.7}$$

(To prove this inequality we first observe that it suffices to consider the case when $t \geq \tau$ and then $|t - a_\tau(t)| = t - \tau$ and we then have

$$4\tau\left(\frac{t^2}{4\tau} - (t-\tau)\right) = (t-2\tau)^2 \geq 0 .)$$

We can now complete the proof of the main lemma as follows:

We have

$$u = \int A(t) \otimes B(t)dt = \int \left(A_\tau(t) + A^\tau(t) \otimes \left(B_\tau(t) + B^\tau(t)\right)dt = \right.$$

$$= \int A_\tau(t) \otimes B_\tau(t)dt + \int A^\tau(t) \otimes B_\tau(t)dt + \int A(t) \otimes B^\tau(t)dt =$$

$$= v + w_1 + w_2 .$$

We then have

$$p_V(v) \leq \int ||A_\tau(t)|| \cdot ||B_\tau(t)||dt \leq \tau^2$$

$$p_U(w_1) \leq || \int A^\tau(t)^2 dt||^{1/2} \cdot || \int B_\tau(t)^2 dt||^{1/2} \leq \left(\frac{3}{16\tau^2}\right)^{1/2} ,$$

by (2.1.5) and (2.1.6), and likewise by (2.1.6) $p_U(w_2) \leq \frac{3}{16\tau^2}^{1/2}$.

We define of course $w = w_1 + w_2$ and choosing $\tau = (\frac{27}{16})^{1/2}$ we have $p_V(v) = \frac{27}{16}$ and $p_U(w) = 2 \cdot \frac{1}{3} = \frac{2}{3}$.

This completes the proof in the special case $A = B(H)$, $\mathbb{B} = \mathbb{B}(K)$. To remove this assumption we simply observe that for every t, the operator $A_\tau(t)$ $\left(B_\tau(t)\right)$ belongs to the C*-algebra generated by the set $\{a_i\}_{i=1}^N$ $(\{b_i\}_{i=1}^N)$. This completes the proof of the main lemma and thereby of the theorem.

<u>Concluding remark</u>. As stated in the corollary of the theorem it follows that the natural map from the projective tensor product $\overline{\mathbb{C}}_V$ to the space $\overline{\mathbb{C}}_U$ is surjective. The proof does not prove, however, that the map is injective. In order to use a decomposition method to prove injectivity the choice of the significant part (corresponding to A_τ) and the small part (corresponding to A^τ) have to be made so that $A^\tau(t)$ $\left(B^\tau(t)\right)$ is orthogonal to both B_τ and B^τ (A_τ and A^τ). In the commutative case, the proof based on Paley's theorem for trigonometric lacunary series, gives a decomposition with these properties. (This fact is pointed out by Fournier [2]).

3. Whether Pisier's theorem about maps from a C*-algebra into a Banach space of cotype 2 can be deduced from the result about bilinear forms on C*-algebras I don't know. On the other hand, it turns out that this result can also be proved by the same method. Towards this we shall need the following two definitions,

Definition (3.1). A Banach space E is said to be of cotype 2 if there exists a constant C such that for every finite subset $\{e_i\} \subset E$, one has

$$(\Sigma \|x_i\|^2)^{1/2} \leq C(\int \|\Sigma\, r_i(t)e_i\|^2 dt)^{1/2} ,$$

where $\{r_i\}$ denotes the set of Rademacher functions on $[0,1]$. The smallest such constant is denoted C_E.

Definition (3.2). Let u be an operator from a C*-algebra A into a Banach space E. u is said to be q - C*-summing, $1 \leq q$ if there exists a constant C, such that for any finite subset $\{a_i\} \subset A_h$ (i.e. of Hermitian elements) one has

$$(\Sigma \|u(a_i)\|^q)^{1/q} \leq C \cdot \|(\Sigma |a_i|^2)^{2/q}\|^{1/q} .$$

The smallest such constant is denoted $C_q(u)$.

This definition is due to Pisier [10].

We shall prove the following

Theorem 2 (Pisier). Let E be a Banach space of cotype 2, let A_h be the self-adjoint part of a C*-algebra and let u be a linear operator from A_h to E. Suppose further that u is 2 - C*-summing and that $C_2(u) = 1$. Then

$$\|u\| \geq (\sqrt{3} \cdot C_E^2)^{-1} .$$

Proof. Let $\varepsilon > 0$ be given. By assumption there exists a finite set $\{a_i\} \subset A_h$ such that

$$(\Sigma \|u(a_i)\|^2 > 1 - \varepsilon ,$$

while $\|\Sigma a_i^2\| \leq 1$. As in the proof of the main lemma we denote $A = A(t) = \Sigma a_i r_i(t)$ and we also define $A_\tau = A_\tau(t)$ and $A^\tau = A^\tau(t)$ as before. We then have

$$1 - \varepsilon \leq \Sigma \|u(a_i)\|^2 \leq C_E^2 \cdot \left(\int \|(\Sigma u(a_i) r_i(t))\|^2 dt\right) =$$

$$= C_E^2 \cdot \left(\int \|u(\Sigma a_i r_i(t))\|^2 dt\right) = C_E^2 \cdot \left(\int \|u(A_\tau(t) + A^\tau(t))\|^2 dt\right) \leq$$

$$\leq 2 \cdot C_E^2 \cdot \left(\int \|u(A_\tau(t))\|^2 dt + \int \|u(A^\tau(t))\|^2 dt\right) .$$

Hence

$$\int \|u(A_\tau(t))\|^2 dt \geq \frac{1-\varepsilon}{2 \cdot C_E^2} - \int \|u(A^\tau(t))\|^2 dt .$$

Now all integrals are actually finite sums, and from the assumption about the $2 - C^*$=summing norm of u it follows that

$$\int \|u(A^\tau(t))\|^2 dt \leq \|\int A^\tau(t)^2 dt\| .$$

By Lemma (2.3) we have again

$$\|\int A^\tau(t)^2 dt\| \leq \frac{1}{16\tau^2} \cdot \|\int A(t)^4 dt\| \leq \frac{3}{16\tau^2}$$

Choosing $\tau = C_E \cdot \sqrt{3/4}$, we have

$$\int \|u(A_\tau(t))\|^2 dt \geq \frac{1-2\varepsilon}{4\, C_E^2}$$

and since $\|A_\tau(t)\| < \tau$ we have

$$\|u\|^2 \geq \frac{1-2\varepsilon}{3\, C_E^4} .$$

Since $\varepsilon > 0$ is arbitrary, the theorem follows.

Acknowledgement. Part of the research for this paper was done while the author was visiting the mathematical department of York University, Toronto, Ont., and it is a pleasure to express my gratitude for the hospitality I enjoyed. I am also grateful to Yngve Domar who pointed out a simplification of the proof, thereby allowing an improvement of the estimate for the constant from just below 7, to just above 5.

References

[1] R. C. Blei, A uniformity property for $\Lambda(2)$ sets and Grothendieck's inequality. Symposia Math. Vol. XXII (1977) 321-336.

[2] J. F. Fournier, On a theorem of Paley and the Littlewood conjecture. Arkiv för Matematik 17 (1979) 199-216.

[3] A. Grothendieck, Résumé de la théorie métrique des produits tensoriels topologiques. Bol. Soc. Mat. Brasil, Sao Paulo, 8 (1956) 1-79.

[4] J. L. Krivine, Théorèmes de factorisations dans les espaces réticulés. Sém. Maurey-Schwarz 73-74, exp. XXII - XXIII.

[5] J. Lindenstrass and A. Pelczynski, Absolutely p-summing operators in L_p-spaces and their applications. Studia Math. 29 (1968) 275-326.

[6] J. Lindenstrass and L. Tzafriri, Classical Banach spaces. Springer, Berlin-Heidelberg-New York 1977.

[7] M. A. Naimark, Normed rings. P. Noordhoff N.V. Groningen 1964.

[8] B. Maurey, Une nouvelle démonstration d'un theorème de Grothendieck. Sém. Maurey-Schwarz 72-73, exp. XXII.

[9] A. Pelczynski, Banach spaces of analytic functions and absolutely summing operators. C.B.M.S. Regional Conference Series in Mathematics, 30, 1977.

[10] G. Pisier, Grothendieck's theorem for non-cummutative C*-algebras with an appendix on Grothendieck's constants. J. Funct. Anal. 29 (1978) 397-415.

Appendix (to "A simple-minded..)

A note on Haagerups theorem

1. Since this paper was written Uffe Haagerup [2] has removed the condition on approximation from the non-cummutative Grothendieck inequality. In his beautiful paper Haagerup not only proves the inequality in full generality, but he also obtains the best constant in the non-commutative case. Nevertheless, his paper is not easy to read, and therefore a simplified version of the proof might be of interest to those interested in the inequality. Let us say that Haagerup's proof consists of a qualitative part where the inequality is proved with some constant and a quantitative part where the best constant is obtained. In this note we have no contribution to make about the quantitative part, but we do have something to say about the qualitative part. In fact, what we want to point out is that already when Pisier's proof appeared it would have been rather easy to complete the proof by using known results. The crucial observation made by Haagerup was that it is possible to prove without assuming the approximation property - that if β is a bilinear form on $A \times B$ (where A and B are unital C*-algebras), such that $\beta(1,1) = \|\beta\| = 1$, then β is (in the terminology of Pisier) 4 - C*-summing. This fact is the only idea in the proof that was not realized when Pisier's paper appeared. It is also interesting to observe that this fact is proved by reducing the non-cummutative case to the classical commutative case. (We shall use the notations introduced in the previous part of the paper.)

2. We shall here give a "new" proof of the following theorem, due to Haagerup.

Theorem 2. (Haagerup)

Let A and B be C*-algebras and let β be a bilinear form on $A \times B$. Let further $\{a_i\}_{i=1}^n \subset A$ and let $\{b_i\}_{i=1}^n \subset B$ be such that

$||\Sigma (a_i^* a + a_i a_i^*)|| \leq 1$ and $||\Sigma (b_i^* b_i + b_i b_i^*)|| \leq 1$. Then

$\Sigma \beta(a_i, b_i) \leq \frac{81}{8} \cdot ||\beta||$.

Remark. This means that the operator $T : A \to B^*$ defined by $\langle T_a, b\rangle = \beta(a,b)$ is $2-C^*$-summing and then (as was proved by Pisier) there exist states p on A and q on B such that

$|\beta(a,b)| \leq \frac{81}{8} \cdot ||\beta|| \cdot p(a^*a + aa^*)^{1/2} \, q(b^*b + bb^*)^{1/2}$.

To prove the theorem we shall need four lemmas.

Lemma 1 (Bohnenblust-Karlin [1]). Let A be a unital Banach algebra, and let S be the state space of A, i.e.

$S = \{s \mid s \in A', \ ||S|| < \langle s, 1_A\rangle = 1\}$. Let further $a \in A$, and let

$$u(a) = \sup_{S \in S} |\langle S_1 a\rangle| .$$

Then $u(a) \geq \frac{1}{e} \cdot ||a||_A$.

Lemma 2 (Maurey-Pisier). Let A be a C^*-algebra and let E be a Banach space of cotype 2. Let further $T : A \to E$ be $4-C^*$-summing. Then T is $2-C^*$-summing and furthermore

$$C_2(u) \leq 7 \cdot C_E \cdot C_4(T) .$$

(Remark: This is an easy consequence of Maurey's generalized Khintchine inequality for Banach spaces of cotype 2, combined with Pisier's inequality (Lemma (2.2) above).)

Lemma 3 (Tomczak-Jaegermann [4]). Let E be the dual space of a C^*-algebra. Then E is a Banach space of cotype 2, and $C_E \leq \sqrt{2e}$.

Lemma 4. Let A and B be unital C*-algebras and let β be a bilinear form on $A \times B$ such that

$$\beta(1_A, 1_B) = 1$$

$\|\beta\| = \beta(1_A, 1_B) = 1.$

Let $T_\beta : A \to B'$ be defined by $\langle T_\beta(a), b\rangle = \beta(a,b)$. Then T_β is 4 - C*-summing and $C_4(T_\beta) \leq 6$.

Lemma 3 is proved by an argument using Riesz-products in [4] and will not be proved here. Lemma 4 will be proved below. The proof will be based on an inequality for bilinear forms on commutative C*-algebras, that was used by the author in another context in [3].

Given the above lemmas we shall now give the

Proof of theorem 2. (We shall assume that A and B are both unital C*-algebras. This is essentially a technical assumption which can easily be removed by standard arguments.)

Let $\overline{C}_V$ be the projective tensor product $A \otimes B$, and let $\overline{C}_U$ be the completion of the algebraic tensor product $A \otimes B$, equipped with the norm given by the Minkowski functional of the convex set U where $U = \{u \in A \otimes B \mid$ there exists $n \in Z^+$ and $\{a_i\} \subset A$, $\{b_i\}_{i=1}^n \subset B$, such that $\|\Sigma (a_i^* a_i + a_i a_i^*)\|_A \leq 1$ and $\|\beta (b_i^* b_i + b_i b_i^*)\|_B \leq 1$, and $u = \sum_{i=1}^n a_i \otimes b_i$.

It follows from theorem 1 that if $h : \overline{C}_V \to \overline{C}_U$ is given by the continuous extension of the identity map on $A \otimes B$, then h is surjective. We shall prove that h is an isomorphism, so we have to prove that it is injective. To do this it suffices to prove that for every $u \in \overline{C}_V$, there exists $\beta \in (\overline{C}_U)'$, such that $\langle\beta, h(u)\rangle \neq 0$.

However, the projective tensor product of unital Banach algebras is again a unital Banach algebra, so $\overline{C}_V$ is therefore a unital Banach algebra. By the Bohnenblust-Karlin theorem (lemma 1 above), there exists therefore a state $S \in C_V'$, such that

$$|\langle S,u\rangle| \geq \frac{1}{e} \cdot \|u\|_{C_V} .$$

It suffices therefore to prove that if S is a state on $\overline{C}_V$, then $S \in (\overline{C}_U)'$. However, S is a state if the corresponding bilinear form β_S defined by $\beta_S(a,b) = \langle S, a \otimes b\rangle$ has the property that $\beta_S(1_A,1_B) = \|\beta_S\| = 1$. Therefore by lemma 4, the linear map $T_S: a \to B'$ defined by

$$\langle T_S(a),b\rangle = \langle s, a \otimes b\rangle$$

is 4 - C*-summing.

By Tomczak-Jaegermann's theorem (lemma 3), B' is of cotype 2, so by the Maurey-Pisier inequality (lemma 2) T_S is 2 - C*-summing and therefore $S \in (\overline{C}_U)'$ and

$$\|S\|_{(\overline{C}_U)'} \leq 7 \cdot C_{B'} \cdot C_4(T_S)$$
$$\leq 7 \cdot \sqrt{2e} \cdot 6$$

It follows that if $u \in \overline{C}_V$, then $\|h(u)\|_{\overline{C}_u} \geq k \cdot \|u\|_{\overline{C}_V}$ where

$$k = (42 \cdot \sqrt{2e})^{-1} \geq \frac{1}{100}$$

This proves that the map h is injective, and this completes the proof of theorem 2. Note that $\|h^{-1}\| \leq 81/8$.

3. It remains to prove lemma 4. For this we shall need three sublemmas, and we want to point out that these three sublemmas are related to C(X)-spaces, while non-commutative C*-algebras play no role.

<u>Sublemma 1</u> [3]. Let X be a compact Haasdorff space and let $\mu,\nu \in M(X)$. Then

$$\sup_{0\leq\theta\leq 2\pi} \|\mu+e^{i\theta}\nu\| \geq (\|\mu\|^2 + \frac{4}{\pi^2}\|\nu\|^2)^{1/2} .$$

This inequality was proved in [3] but since that paper is probably not widely known, we shall give an outline of the proof.

We start by observing that since both μ and ν are absolutely continuous with respect to e.g. the measure $\lambda = (|\mu|+|\nu|)$, we may assume that $\mu = fd\lambda$, $\nu = gd\lambda$ so we shall prove that if $f,g \in L^2(X,d\lambda)$ then

$$\sup_{0\leq\theta\leq 2\pi} \int |f + e^{i\theta}g|d\lambda \geq (\|f\|_1^2 + \frac{4}{\pi^2} \|g\|_1^2)^{1/2}.$$

Let us start from the left.

$$\sup_{0\leq\theta\leq 2\pi} \int_X |f + e^{i\theta}g|d\lambda \geq \frac{1}{2\pi}\int_0^{2\pi} \int |f(x) + e^{i\theta}g(x)|d\lambda(x)d\theta$$

$$= \int_X \frac{1}{2\pi}\int_0^{2\pi} |f(x) + e^{i\theta}g(x)|d\theta d\lambda(x) \quad \text{(as is easily seen)}$$

$$= \int_X \frac{1}{2\pi}\int_0^{2\pi} ||f(x)| + e^{i\theta} |g(x)||d\theta d\lambda(x)$$

$$= \int_X \frac{1}{\pi}\int_0^{\pi} ||f(x)| + e^{i\theta}|g(x)|| d\theta d\lambda(x)$$

$$\geq | \int_X \frac{1}{\pi}\int_0^{\pi} (|f(x)| + e^{i\theta}|g(x)|| d\theta d\lambda(x) |$$

$$\leq | \|f\|_1 + \frac{2i}{\pi} \|g\|_1| \geq (\|f\|^2 + \frac{4}{\pi^2} \|g\|^2)^{1/2}$$

and this proves the sublemma.

<u>Sublemma 2</u>. Let X and Y be compact Hausdorff spaces and let β be a bilinear form on $C(X)\times C(Y)$ such that $\|\beta\| = \beta(1,1) = 1$.

Let μ be the probability measure on X defined by $\int_X f\,d\mu = \beta(f,1)$. Let finally $T_\beta : C(X) \to M(Y)$ be defined by $\langle T_\beta(f), g\rangle = \beta(f,g)$ and let $f \in C(X)$. Then

$$\|T_\beta(f)\| \leq \frac{\pi}{\sqrt{2}}\left(\int |f|\,d\mu\right)^{1/2} \cdot \|f\|_\infty^{1/2}$$

Proof. Let $g = \frac{f}{\|f\|}$ and let $h = 1 - |g|$. Let further $\nu_g \in M(Y)$ and $\nu_h \in M(Y)$ be defined by $\int_Y \phi\,d\nu_g = \beta(g,\phi)$ resp. $\int_Y \phi\,d\nu_h = \beta(h,\phi)$.

Now

$$\|\nu_h + e^{i\theta}\nu_g\| = \|T_\beta(h + e^{i\theta}g)\| \leq \|\beta\| \cdot \|h + e^{i\theta}g\| = 1.$$

However by sublemma 1

$$\sup \|\nu_h + e^{i\theta}\nu_g\| \geq \left(\|\nu_h\|^2 + \frac{4}{\pi^2}\|\nu_g\|^2\right)^{1/2}.$$

Let $\|f\|_\infty = m$ and let $a = \int |f|\,d\mu$.

Then

$$\|\nu_h\| \geq \int 1\,d\nu_h = \beta(h,1)$$

$$= \int_X (1 - |g|)\,d\mu = 1 - \frac{a}{m}.$$

Therefore

$$\left(1 - \frac{a}{m}\right)^2 + \frac{4}{\pi^2}\|\nu_g\|^2 \leq 1$$

so $$\|\nu_g\| \leq \frac{\pi}{2}\left(1 - \left(1 - \frac{a}{m}\right)^2\right)^{1/2} \leq \frac{\pi}{2}\left(2\frac{a}{m} - \frac{a^2}{m^2}\right)^{1/2}$$

$$\leq \frac{\pi}{\sqrt{2}}\left(\frac{a}{m}\right)^{1/2}.$$

Therefore

$$\|T_\beta(f)\| = m\cdot\|T_\beta(g)\| \leq \frac{\pi}{\sqrt{2}}\cdot\|f\|_\infty^{1/2}\cdot\left(\int|f|d\mu\right)^{1/2},$$

and this proves the sublemma.

<u>Sublemma 3</u> (Standard interpolation theory). Let E be a Banach space, let X be a compact Hausdorff space, let μ be a probability measure on X and let $T: C(X) \to E$ be an operator such that for $f \in C(X)$

$$\|Tf\| \leq C\cdot\|f\|_\infty^{1/2}\cdot\|f\|_{L^1(\mu)}^{1/2}$$

Then T extends by continuity to $L^4(d\mu)$ and

$$\|Tf\| \leq 2\cdot\|T\|^{1/2}\cdot C^{1/2}\cdot\|f\|_{L^4}$$

(Remark: It can be proved that T is in fact continuous on the Lorentz space $L^{2,1}(d\mu)$.)

<u>Proof</u>. Let $f \in C(X)$ and suppose that $\int|f|^4 d\mu = 1$.

Let the function $a: C \to C$ be defined by

$$a(z) = \frac{z}{\max(|z|,1)}$$

and let a_τ, $0 < \tau$ be defined by

$$a_\tau(z) = \tau\cdot a\left(\frac{z}{\tau}\right)$$

Let now $m_k = \tau\cdot 2^k$ and let us write

$$f_o = a_\tau(f),$$

$$f_k = a_{n_k}(f) - a_{n_{k-1}}(f), \quad k = 1,2,\ldots, \quad .$$

Then $f = \sum_{k=0}^{N} f_k$ (for some N) so we have

$$Tf = T(\Sigma f_k) = \Sigma\, T(f_k)$$

and hence $\|Tf\| \le \Sigma \|T(f_k)\|$.

We have now

$$\|f_o\|_\infty = \tau \quad \text{and} \quad \|f_k\|_\infty = 2^{k-1}\cdot\tau\,..$$

Therefore

$$\sum_{k=o}^{\infty} \|T f_k\| \le \|T\|\cdot\|f_o\|_\infty + C\cdot\sum_{k=1}^{N} \|f_k\|_\infty^{1/2}\cdot\|f_k\|_1^{1/2}$$

(where $\|\phi\|_1 = \int |\phi|\, d\mu$).

$$\le \tau\cdot\|T\| + C\cdot\sum_{k=1}^{N} (\tau\cdot 2^{k-1})^{1/2}\cdot\|f_k\|_1^{1/2}.$$

Let us denote $b_o = a_\tau$ and $a_{n_k} - a_{n_{k-1}} = b_k$.

We observe then that for $t \ge 0$

$$b_o(t) + \sum_{k=1}^{\infty} 4\cdot(\tau\cdot 2^{k-1})^3\cdot b_k(t) \le \tau + t^4$$

$$\text{so} \quad \sum_{k=1}^{\infty} 4\cdot(\tau\cdot 2^{k-1})^3\cdot b_k(t) \le t^4 .$$

We have therefore

$$\tau\cdot\|T\| + C\cdot\sum_{k=1}^{N} (\tau\cdot 2^{k-1})^{1/2}\cdot\|f_k\|_1^{1/2}$$

$$= \tau\cdot\|T\| + C\cdot\sum_{k=1}^{N} (\tau\cdot 2^{k-1})^{1/2}\big(4(\tau\cdot 2^{k-1})^3\big)^{-\frac{1}{2}}\cdot\big(4(\tau\cdot 2^{k-1})^3\big)^{1/2}\cdot\|f_k\|_1^{1/2}$$

$$\le \tau\cdot\|T\| + C\Big\{\sum_{k=1}^{N} \tau\cdot 2^{k-1}\cdot\big(4\cdot(\tau\cdot 2^{k-1})^3\big)^{-1}\Big\}^{1/2}\Big\{\sum_{k=1}^{N} 4\cdot(\tau\cdot 2^{k-1})^3\|f_k\|_1\Big\}^{1/2}$$

$$\leq \tau\cdot\|T\| + C\cdot\{\sum_{k=1}^{\infty} \frac{1}{\tau^2}\cdot 2^{-2k}\}^{1/2}\{\int |f|^4\,d\mu)^{1/2}$$

$$\leq \tau\cdot\|T\| + C\cdot \frac{1}{\tau\cdot\sqrt{3}} \leq \tau\cdot\|T\| + \frac{C}{\tau}.$$

Choosing $\tau = \left(\frac{C}{\tau}\right)^{1/2}$ we get then

$$\|Tf\| \leq \Sigma\,\|T f_k\| \leq 2\cdot C^{1/2}\cdot\|T\|^{1/2}$$

and since $\|f\|_4 = 1$, the sublemma follows.

4. We can now give the proof of Lemma 4.

Proof. We are given a bilinear form β on $A \times B$ such that $\beta(1,1) = \|\beta\| = 1$, and we want to prove that the associated linear map $T_\beta : A \to B'$ is $4 - C^*$-summing, i.e. that there exists C such that if $\{a_i\}_{i=1}^n \subset A_h$ then

$$(\Sigma\,\|T_\beta a_i\|^4)^{1/4} \leq C\,\|(\Sigma\, a_i^4)^{1/4}\|.$$

To this end, we define a state p on A by

$$p(a) = \beta(a,1).$$

We shall prove that if $a \in A_h$, then

$$\|T_\beta(a)\|^4 \leq C^4\,p(a^4);$$

for then clearly

$$\Sigma\,\|T_\beta(a_i)\|^4 \leq C^4\cdot\Sigma\,p(a_i^4) = C^4\,p(\Sigma\, a_i^4)$$

$$\leq C^4\cdot\|\Sigma\, a_i^4\|.$$

Now

$$\|T_\beta(a)\| > \sup\{|\beta(a,b)| \;\big|\; b \in \mathcal{B},\ \|b\| \le 1\}$$

$$\le 2\sup\{|\beta(a,b)| \;\; b \in \mathcal{B}_h,\ \|b\| \le 1\},$$

so it suffices to prove that

$$|\beta(a,b)| \le C \cdot p(a^4)^{1/4} \cdot \|b\|$$

if a and b are self-adjoint.

Let therefore $a \in A$ ($b \in \mathcal{B}$) be self-adjoint and let $\tilde{A} \subset A$ ($\tilde{\mathcal{B}} \subset \mathcal{B}$) be the subalgebra generated by a and 1_A (b and $1_{\mathcal{B}}$). Then $\tilde{A}$ ($\tilde{\mathcal{B}}$) is a commutative C^*=algebra so we have $\tilde{A}$ isomorphic to $C(X)$ ($\tilde{\mathcal{B}}$ isomorphic to $C(Y)$) for some compact Hausdorff space $X(Y)$. We consider the restriction $\tilde{\beta}$ of β to $\tilde{A} \times \tilde{\mathcal{B}}$ and we observe that $\tilde{\beta}$ fulfills the conditions for sublemma 2 so that by sublemma 2

$$\|T_{\tilde{\beta}}(f)\| \le \frac{\pi}{\sqrt{2}} \left(\int |\hat{a}(x)|d\mu(x)\right)^{1/2} \cdot \|\hat{a}\|_\infty^{1/2}$$

where $\hat{a}(x)$ is the Gelfand transform of a (in $C(X)$), and where $\int\hat{a}(x)d\mu(x) = \beta(a,1) = \beta(a,1) = p(a)$.

By sublemma 3 we have therefore

$$\|T_{\tilde{\beta}}(a)\| \le 2\left(\frac{\sqrt{\pi}}{\sqrt{2}}\right)^{1/2}\left(\int|\hat{a}(x)|^4\, d\mu(x)\right)^{1/4} = 2^{7/4}\pi^{1/2}\, p(a^4)^{1/4},$$

and hence if a and b are self-adjoint

$$|\beta(a,b)| \le C\cdot p(a^4)^{1/4} \cdot \|b\|$$

where e.g. $C = 2^{7/4} \cdot \pi^{1/2} \le 6$, and this proves lemma 4, which completes the proof of Haagerup's theorem.

References for the Appendix

[1] Bohnenblust, H. F., and Karlin, S., Geometrical properties of the unit sphere of Banach algebras. Ann. of Math. 62 (1955) 217-229.

[2] Haagerup, Uffe, The Grothendieck inequality for bilinear forms on C*-algebras. Preprint, Odense University.

[3] Kaijser, S., Representations of tensor algebras as quotients of group algebras. Arkiv f. Mat. 10 (1972) 107-141.

[4] Tomczak-Jaegermann., On the moduli of smoothness and convexity and the Rademacher averages of the trace classes S_p $(1 \leq p < \infty)$. Studia Math. 50 (1974) 163-182.

THE BEHAVIOR OF POWER SERIES ON THEIR CIRCLE OF CONVERGENCE

T. W. Körner

§1 Introduction

One of the first things a student learns in complex variable theory is the existence of a radius of convergence R for each power series.

Theorem 1.1. If $a_o, a_1, \ldots \in \mathbf{C}$ then either $\sum_{n=0}^{\infty} a_n z^n$ converges for all z or there exists an $R \geq 0$ such that $\sum_{n=0}^{\infty} a_n z^n$ converges for $|z| < R$ and diverges for all $|z| > R$.

Proof. Omitted. □

She is then warned that Theorem 1.1 says nothing about the behavior of $\sum_{n=0}^{\infty} a_n z^n$ when $|z| = R$.

Lemma 1.2. (i) $\sum_{n=0}^{\infty} n\, z^n$ has radius of convergence 1 and diverges unboundedly for all $|z| = 1$.

(ii) $\sum_{n=1}^{\infty} n^{-2} z^n$ has radius of convergence 1 and converges for all $|z| = 1$.

(iii) $\sum_{n=1}^{\infty} n^{-1} z^n$ has radius of convergence 1, converges for all $|z| = 1$, $z \neq 1$ and diverges unboundedly for $z = 1$.

Proof. Parts (i) and (ii) are obvious. We include a proof of (iii) simply for the pleasure of using Abel summation. Set $S_r = \sum_{n=0}^{r} z^n$ so that $S_r = (1 - z^{r+1})/(1 - z)$ $[z \neq 1]$ and $|S_r| \leq 2/(1 - |z|)$. Then, if $|z| \neq 1$ and $z \neq 1$,

$$\left|\sum_{n+N}^{M} n^{-1} z^n\right| = \left|\sum_{n=N}^{M} n^{-1} (s_n - s_{n-1})\right|$$

$$= \left|\sum_{n=N}^{M} \left(n^{-1} - (n+1)^{-1}\right) s_n + (M+1)^{-1} s_M\right|$$

$$\leq \sum_{n=N}^{M} \left(n^{-1} - (n+1)^{-1}\right) |s_n| + (M+1)^{-1} |s_M|$$

$$\leq 2(1-|z|)^{-1}\left(\sum_{n=N}^{M} (n^{-1} - (n+1)^{-1}) + (M+1)^{-1}\right)$$

$$= 2N^{-1}(1-|z|)^{-1} \to 0 \quad \text{as} \quad N \to \infty .$$

Thus by the general principle of convergence $\sum_{n=1}^{\infty} n^{-1} z^n$ converges for $|z| = 1$, $z \neq 1$. Since Σn^{-1} diverges we are done. □

To the examples above we may add others like the following one.

__Theorem 1.3__. There exist $a_0, a_1, a_2, \ldots \in \mathbb{C}$ such that $\sum_{n=0} a_n z^n$ converges for $|z| = 1$, $z \neq 1$ and diverges boundedly for $z = 1$.

Our construction uses Fejer's kernel

$$\sigma_n(t) = \sum_{r=-n}^{n} \frac{(n+1-|r|)}{n+1} \exp irt.$$

__Lemma 1.4__. (i) $\sigma_n(t) = \frac{1}{n+1}\left(\frac{\sin\ (n+1)t/2}{\sin\ t/2}\right)^2 \qquad t \not\equiv 0 \mod 2\pi$

(ii) $\sigma_n(0) = \sum_{r=-n}^{n} \frac{(n+1-|r|)}{n+1} = n+1$

(iii) $\sigma_n(t) \to 0$ uniformly for $\pi \geq |t| \geq \varepsilon \qquad [\varepsilon > 0]$

__Proof__. (i) $(n+1)\sigma_n(t) = \left(\sum_{r=0}^{n} \exp(i(2r-n)t/2)\right)^2$

$$= (\exp -int)\left(\sum_{r=0}^{n} \exp irt\right)^2$$

$$= (\exp -int)\left(\frac{1-\exp\ i(n+1)t}{1-\exp it}\right)^2$$

$$= \left(\frac{\exp i(n+1)t/2-\exp-i(n+1)t/2}{\exp it/2-\exp-it/2}\right)^2$$

$$= \left(\frac{\sin(n+1)t/2}{\sin t/2}\right)^2$$

(ii) Obvious.

(iii) Observe that, by (i), $|\sigma_n(t)| \leq \frac{1}{n+1}\left(\frac{1}{\sin\ /2}\right)^2 \to 0$ as $n \to \infty$ for $\pi \geq |t| \geq \varepsilon$. □

Lemma 1.5. (i) Given $\varepsilon > 0$ we can find $N \geq 1$ and $a_o, a_1, \ldots, a_{N-1} \geq 0$ such that

(α) $\sum_{j=o}^{N-1} a_j = 1$

(β) $|\sum_{j=o}^{N-1} a_j z^j| \leq \varepsilon$ for $|z| = 1$, $|z-1| \geq \varepsilon$.

(ii) Given $\eta > 0$ we can find $M \geq 1$ and $b_o, b_1, \ldots, b_M \geq 0$ such that

(α) $\sum_{j=o}^{M-1} b_j = 1$

(β) $|\sum_{j=m(1)}^{m(2)} b_j z^j| \leq \eta$ for $|z| = 1$, $|z-1| \geq \varepsilon$

and all $0 \leq m(1) \leq m(2) \leq M-1$.

Proof. (i) Let $N = 2n+1$ and

$$\sum_{j=o}^{N-1} a_j z^j = \frac{z^n}{(n+1)} \sum_{r=-n}^{n} \frac{(n+1-|r|)}{n+1} z^r .$$

Then (α) is automatic and (β) follows if we take n large enough and observe that $\sum_{j=o}^{N-1} a_j (\exp it)^j = (n+1)^{-1}\sigma_n(t)$ so that Lemma 1.4 (iii) is applicable.

(ii) Choose an integer $P \geq 4\eta^{-1}$ and set $\varepsilon = \eta/4$. Choose $a_o, a_1, \ldots, a_{N-1}$ as in (i) and take $M = PN$,

$$\sum_{j=o}^{M-1} b_j z^j = \sum_{r=o}^{P-1} P^{-1} z^{rN} \sum_{k=o}^{N-1} a_k z^k$$

so that $b_{rN+k} = P^{-1} a_k$ $[0 \le r \le P-1,\ 0 \le k \le N-1]$. Condition (α) is automatically satisfied and to check (β) we proceed as follows. If $0 \le m \le M-1$ then we can find $0 \le n \le P-1$ and $0 \le \ell \le N-1$ such that $m = n + \ell$ and so, if $|z| = 1$

$$\begin{aligned}
|\sum_{j=o}^{m} b_j z^j| &= |\sum_{r=o}^{n-1} P^{-1} z^{rN} \sum_{k=o}^{N-1} a_k z_k + P^{-1} z^{nN} \sum_{k=o}^{\ell} a_k z^k| \\
&\le \sum_{r=o}^{n-1} P^{-1} |\sum_{k=o}^{N-1} a_k z^k| + P^{-1} \sum_{k=o}^{\ell} a_k \\
&\le \sum_{r=o}^{n-1} \varepsilon P^{-1} + P^{-1} \\
&\le \varepsilon + P^{-1} \le \eta/2
\end{aligned}$$

for all z with $|z-1| \ge \varepsilon$ and so for all z with $|z-1| \ge \eta$. It follows that

$$\begin{aligned}
|\sum_{j=m(1)}^{m(2)} b_j z^j| &= |\sum_{j=o}^{m(2)} b_j z^j - \sum_{j=o}^{m(1)-1} b_j z^j| \\
&\le |\sum_{j=o}^{m(2)} b_j z^j| + |\sum_{j=o}^{m(1)-1} b_j z^j| \\
&\le \eta/2 + \eta/2 = \eta
\end{aligned}$$

for all z with $|z| = 1$ and $|z-1| \ge \eta$. □

<u>Proof of Theorem 1.3</u>. By Lemma 1.5 we can find $M(1), M(2), \ldots$ and $b_{10}, b_{11}, \ldots, b_{1M(1)}, b_{20}, b_{21}, \ldots, b_{2M(2)}, \ldots$ such that $b_{pj} \ge 0$ $[0 \le j \le M(p)-1,\ 1 \le p]$ and

$$(\alpha)_p \qquad \sum_{j=o}^{M(p)-1} b_{pj} = 1$$

$$(\beta)_p \qquad \left|\sum_{j=m(1)}^{m(2)} b_{pj} z^j\right| \le 2^{-p} \quad \text{for} \quad |z|,\ |z-1| \ge 2^{-p}$$

and all $0 \le m(1) \le m(2) \le M(p) - 1$.

Set $N(p) = M(1) + M(2) + \ldots + M(p-1)$ for $p \ge 2$, $N(1) = 0$ and define a_j by the (formal) relation

$$\sum_{j=0}^{\infty} a_j z^j = \sum_{p=1}^{\infty} (-1)^{p+1} z^{N(p)} \sum_{k=0}^{m(p)-1} b_{pk} z^k$$

so that

$$a_{M(1)+M(2)+\ldots\ldots+M(p-1)+k} = a_{N(p)+k} = (-1)^{p+1} b_{pk}$$

for $0 \le k \le M(p) - 1$.

Suppose $|z| = 1$ and $z \neq 1$. Then $|z - 1| > 0$ and so given $\varepsilon > 0$ we can find a q such that $|z-1| < 2^{-q-1}$ and if $\varepsilon < 2^{-q+3}$ if $j(2) > p(1) > N(q)$ then we can find $p(2) \ge p(1) \ge q$ and $k(1)$, $k(2)$ with $M(p(u)) - 1 \ge k(u) \ge 0$ and $j(u) = N(p(u)) + k(u)$ $[u = 1,2]$. We then have

$$\left|\sum_{j=j(1)}^{j(2)} a_j z^j\right| \le \left|\sum_{j=n(p(1))}^{j(2)} a_j z^j\right| + \left|\sum_{j=N(p(1))}^{j(1)} a_j z^j\right|$$

$$\le \sum_{p=p(1)}^{p(2)-1} \left|\sum_{j=N(p)}^{N(p+1)-1} a_j z^j\right| + \left|\sum_{j=N(p(2))}^{j(2)} a_j z^j\right| + \left|\sum_{j=N(p(1))}^{j(1)} a_j z^j\right|$$

$$= \sum_{p=p(1)}^{p(2)-1} \left|\sum_{j=0}^{M(p)} b_{pj} z^j\right| + \left|\sum_{j=0}^{k(2)} b_{p(2)j}\right| + \left|\sum_{j=0}^{k(1)} b_{p(1)j}\right|$$

$$\le \sum_{p=p(1)}^{p(2)-1} 2^{-p} + 2^{-p(2)} + 2^{-p(1)} \le 2^{-p(1)+1} + 2^{-p(2)} + 2^{-p(1)}$$

$$< 2^{-q+3} < \varepsilon .$$

Thus by the general principle of convergence $\sum_{j=0}^{\infty} a_j z^j$ converges for all $|z| = 1$, $z \neq 1$.

Again since

$$\sum_{j=0}^{N(p)+k} a_j = \sum_{r=1}^{p-1} \sum_{j=0}^{M(r)-1} a_{N(r)+j} + \sum_{j=0}^{k} a_{N(p)+j}$$

$$= \sum_{r=1}^{p-1} (-1)^{r+1} \sum_{j=0}^{M(r)-1} b_{rj} + (-1)^{p+1} \sum_{j=0}^{k} b_{pj}$$

$$= \sum_{r=1}^{p-1} (-1)^{r+1} + (-1)^{p+1} \sum_{j=1}^{k} b_{pj}$$

it is easy to see that $\sum_{j=0}^{m} a_j$ oscillates between 0 and 1 as $m \to \infty$ so $\sum_{j=0}^{\infty} a_j z^j$ diverges boundedly for $z = 1$. □

Armed with these examples the lecturer may then add that practically anything can happen on the circle of convergence. The object of this essay is to consider this statement more closely.

§2 Topological Restrictions

There is in fact an obvious restriction on the type of subset of the circle of convergence on which $\sum_{n=0}^{\infty} a_n z^n$ can diverge. This restriction follows from the fact that $S_N(z) = \sum_{n=0}^{\infty} a_n z^n$ is a continuous function and the set on which a sequence of continuous functions can diverge must have a simple topological type.

In this section we work in a metric space (X,d). If F is a closed set we write

$$d(x,F) = \inf \{d(x,y) : y \in F\} .$$

Definition 2.1. If $\mathcal{C}$ is a collection of sets we write $\mathcal{C}_\sigma$ for the collection of countable unions of sets in $\mathcal{C}$ and $\mathcal{C}_\delta$ for the collection of countable intersections. We write $\mathcal{F}$ for the collection of closed sets and $\mathcal{G}$ for the collection of open sets.

Thus $\mathcal{C}_{\sigma\sigma} = \mathcal{C}_\sigma$, $\mathcal{C}_{\delta\delta} = \mathcal{C}_\delta$, $\mathcal{F}_\delta = \mathcal{F}$, $\mathcal{G}_\sigma = \mathcal{G}$ and

$$\{X \setminus C : C \in \mathcal{C}\}_\sigma = \{X \setminus C : C \in \mathcal{C}_\delta\}$$

$$\{X \setminus C : C \in \mathcal{C}\}_\delta = \{X \setminus C : C \in \mathcal{C}_\sigma\} .$$

We note that (in a metric space) $\mathcal{F} \subseteq \mathcal{G}_\delta$ since if $F \in \mathcal{F}$ then $F_n = \{x : d(x,F) < 1/n\} \in \mathcal{G}$ and $F = \bigcap_{n=1}^{\infty} F_n$. By taking complements $\mathcal{G} \subseteq \mathcal{F}_\sigma$.

<u>Theorem 2.2</u>. Let $f_1, f_2, \ldots$ be a sequence of complex valued continuous functions on (X,d). Then the set of points x where $f_n(x)$ converges is in $\mathcal{F}_{\sigma\delta}$.

<u>Proof</u>. Using the general principle of convergence we have

$$\{x : f_n(x) \text{ converges}\}$$

$$= \{x : \text{given } \varepsilon > 0 \text{ there exists an } M_o(\varepsilon,x) \text{ such that } |f_n(x) - f_m(x)| \le \varepsilon \text{ for } n, m \ge M_o\}$$

$$= \bigcap_{N=1}^{\infty} \{x : \text{there exists an } M_o(N^{-1},x) \text{ such that } |f_n(x) - f_m(x)| \le N^{-1} \text{ for } n, m \ge M_o\}$$

$$= \bigcap_{N=1}^{\infty} \bigcup_{M=1}^{\infty} \{x : |f_n(x) - f_m(x)| \le N^{-1} \text{ for } n, m \ge M\}$$

$$= \bigcap_{N=1}^{\infty} \bigcup_{M=1}^{\infty} \bigcap_{n=M}^{\infty} \bigcap_{m=M}^{\infty} \{x : |f_n(x) - f_m(x)| \le N^{-1}\}$$

$\in \mathcal{F}_{\delta\delta\sigma\delta} = \mathcal{F}_{\sigma\delta}$, as stated. □

However there are no further topological restrictions on the set of points on which a sequence of continuous functions can converge as the following theorem due to Hahn and Sierpinski (independently) shows.

Theorem 2.3. If $E \in F_{\sigma\delta}$ then there exists a sequence of real valued continuous functions $f_1, f_2, \ldots$ with $E = \{x : f_n(x) \text{ converges}\}$.

We shall prove Theorem 2.3 in its complemented form.

Theorem 2.3'. If $E' \in G_{\delta\sigma}$ then there exists a sequence of real valued continuous functions $f_1, f_2, \ldots$ with $E' = \{x : f_n(x) \text{ diverges}\}$.

Proof of Equivalence of Theorems 2.3 and 2.3'. Take $E' = X \setminus E$. □

We build up to Theorem 2.3' step by step stating from the following basic lemma.

Lemma 2.4 (The ripple lemma). If $U \in G$ then we can find a sequence of real valued continuous functions $f_1, f_2, \ldots$ such that

(i) For all $x \in X$, $1 \geq f_n(x) \geq 0$

(ii) For all $x \notin U$, $f_n(x) = 0$

(iii) For all $x \in U$ there exists at least one value of n such that $f_n(x) = 1$

(iv) For all $x \in U$ there exist only finitely many values of n such that $f_n(x) \neq 0$.

Proof. We define a sequence of piecewise linear functions $g_n : \mathbb{R} \to \mathbb{R}$ as follows. The function g_1 is given by $g_1(t) = 1$ for $t \geq 2^{-2}$, $g_1(t) = 0$ for $2^{-2} \geq t$ and the condition g_1 linear on $[2^{-2}, 2^{-1}]$. If $n \geq 2$ the function g_n is given by $g_n(t) = 0$ for $t \geq 2^{-n+1}$, $g_n(t) = 1$ for $2^{-n} \geq t \geq 2^{-n-1}$, $g_n(t) = 0$ for $2^{-n-2} \geq t$ and the conditions g_n linear on $[2^{-n+1}, 2^{-n}]$ and $[2^{-n-2}, 2^{-n-1}]$. Observe that

(i)' For all $t \in \mathbb{R}$, $1 \geq g_n(t) \geq 0$

(ii)' For all $t \leq 0$, $g_n(t) = 0$

(iii)' For $t > 0$ there exists at least one value of n such that $g_n(t) = 1$

(iv)' For all $t > 0$ there exist at most 3 values of n such that $g_n(t) \neq 0$.

Now let $F = X \setminus U$ and set $f_n(x) = g_n(d(x,F))$. Conditions (i), (ii), (iii) and (iv) can be read off directly from conditions (i)', (ii)', (iii)' and (iv)'. □

Remark. The name "Ripple Lemma" seems appropriate since if $f_n(x)$ is the height of the water at x at time n we obtain a ripple spreading outward on the pond U.

Lemma 2.5. If $U_1, U_2, \ldots \varepsilon G$ then we can find a sequence of real valued continuous functions $f_1, f_2, \ldots$ such that

(i) For all $x \varepsilon X$, $1 \geq f_n(x) \geq 0$,

(ii) For all $x \varepsilon \bigcap_{j=1}^{\infty} U_j$ there exist infinitely many values of n such that $f_n(x) = 1$ and infinitely many values of n such that $f_n(x) = 0$,

(iii) For all $x \notin \bigcap_{j=1}^{\infty} U_j$ there exist only finitely many values of n such that $f_n(x) \neq 0$.

Proof. By considering $V_1 = U_1$, $V_2 = U_1 \cap U_2, \ldots, V_n = \bigcap_{J=1}^{n} U_j, \ldots$ if necessary, we may suppose, without loss of generality that $U_1 \supseteq U_2 \supseteq \ldots$. By Lemma 2.4 we can find, for each $m \geq 1$, continuous functions $g_{m1}, g_{m2}, \ldots$ such that

$(i)_m$ For each $x \varepsilon X$, $1 \geq g_{mn}(x) \geq 0$,

$(ii)_m$ For each $x \notin U_m$, $g_{nm}(x) = 0$,

$(iii)_m$ For each $x \in U_m$ there exists at least one value of n such that $g_{nm}(x) = 1$,

$(iv)_m$ For all $x \in U_m$ there exist only finitely many values of n such that $g_{mn}(x) \neq 0$.

We now consider the possible values of $g_{mn}(x)$ as both m and n are allowed to vary freely. Obviously

(i)' For each $x \in X$, $1 \geq g_{mn}(x) \geq 0$.

Now suppose $x \notin \bigcap_{j=1}^{\infty} U_j$. Since we have taken $U_1 \supseteq U_2 \supseteq \ldots$ there exists an m_0 such that $x \notin U_m$ for all $m > m_0$. It follows by $(ii)_m$ that $g_{nm}(x) = 0$ for all $m > m_0$ and all n. On the other hand, by $(iv)_m$ for each value $1 \leq m \leq m_0$, there are only a finite number of values of n for which $g_{mn}(x) \neq 0$. We conclude

(ii)' If $x \notin \bigcap_{j=1}^{\infty} U_j$ then there are only a finite number of ordered pairs (m,n) with $g_{mn}(x) \neq 0$.

If $x \in \bigcap_{j=1}^{\infty} U_j$ we know that for each $m \geq 1$, $x \in U_m$ and so by $(iii)_m$, we can find a $p(m)$ with $g_{mp(m)}(x) = 1$. On the other hand $(iv)_m$ shows that for each m we can find a $g(m)$ such that $g_{mq(m)}(x) = 0$. We conclude

(iii)' If $x \in \bigcap_{j=1}^{\infty} U_j$ then there are an infinite number of ordered pairs (m,n) with $g_{mn}(x) = 1$ and an infinite number of ordered pairs (m,n) with $g_{mn}(x) = 0$. We now take $f_1, f_2, \ldots$ to be some enumeration of the g_{mn} (for example we could take $f_{\frac{1}{2}n(n-1)+r} = g_{n+1-r,r}$ $[1 \leq r \leq n, 1 \leq n]$). Conditions (i), (ii) and (iii) are then just restatements of conditions (i)', (ii)' and (iii)'.

□

Proof of Theorem 2.3'. Suppose that $E' \varepsilon G_{\delta\sigma}$. Then we can find $V_1, V_2, \ldots \varepsilon G_\delta$ such that $E' = \bigcup_{j=1}^{\infty} V_j$. By Lemma 2.5 we can find, for each $m \geq 1$, continuous functions $g_{m1}, g_{m2}, \ldots$ such that

$(i)_m$ For each $x \varepsilon X$, $1 \geq g_{mn}(x) \geq 0$,

$(ii)_m$ For each $x \varepsilon V_m$ there exist infinitely many values of n such that $g_{mn}(x) = 1$ and infinitely many values of n such that $g_{mn}(x) = 0$,

$(iii)_m$ For each $x \notin V_m$ there exist only finitely many values of n such that $g_{mn}(x) \neq 0$.

Now set $h_{mn} = 2^{-m} g_{mn}$. Conditions $(i)_m$, $(ii)_m$, $(iii)_m$ give rise directly to the following conditions.

$(i)_m'$ For each $x \varepsilon X$, $2^{-m} \geq h_{mn}(x) \geq 0$.

$(ii)_m'$ For each $x \varepsilon V_m$ there exist infinitely many values of n such that $h_{mn}(x) = 2^{-m}$ and infinitely many values of n such that $h_{mn}(x) = 0$.

$(iii)_n'$ For each $x \notin V_m$ there exist only finitely many values of n such that $h_{mn}(x) \neq 0$.

We now consider the possible values of $h_{mn}(x)$ as both m and n are allowed to vary freely. Suppose first that $x \varepsilon \bigcup_{j=1}^{\infty} V_j$. Then $x \varepsilon V_k$ for some $k \geq 1$ and, by $(ii)_k'$, $g_{kn}(x)$ takes the values 0 and 2^{-k} infinitely often. We conclude

(A) If $x \varepsilon \bigcup_{j=1}^{\infty} V_j$ there exists an $\eta(x) > 0$ such that there are an infinite number of ordered pairs (m,n) with $h_{mn}(x) \geq \eta$ and an infinite number of ordered pairs (m,n) with $h_{mn}(x) = 0$.

Now suppose that $x \notin \bigcup V_j$ and $\varepsilon > 0$. Choose an $m_o \geq 1$ such that $\varepsilon \geq 2^{-m_o}$. It follows by $(i)_m'$ that $\varepsilon > h_{mn}(x) \geq 0$ for all

$m > m_0$ and all n. On the other hand by $(iii)_m'$, for each time $1 \leq m \leq m_0$, there are only a finite number of values of n for which $h_{mn}(x) \neq 0$ and so, a priori, only a finite number of values of n for which $h_{mn}(x) > \varepsilon$. We conclude

(B)' If $x \notin \bigcup_{j=1}^{\infty} V_j$ and $\varepsilon > 0$ then $\varepsilon > h_{mn}(x) \geq 0$ for all but a finite number of pairs (m,n).

We now take $f_1, f_2, \ldots$ to be some enumeration of the h_{mn} (for example we could take $f_{\frac{1}{2}n(n-1)+r} = h_{n+1-r,r}$ $[1 \leq r \leq n,\ 1 \leq n]$). Conditions (A)' and (B)' can now be restated as follows.

(A) If $x \in \bigcup_{j=1}^{\infty} V_j$ then there exists an $\eta > 0$ and two sequences $n(1) < n(2) < n(3) < \ldots$, $m(1) < m(2) < \ldots$ such that $f_{n(k)}(x) \geq \eta$, $f_{m(k)}(x) = 0$.

(B) If $x \notin \bigcup_{j=1}^{\infty} V_j$ then given $\varepsilon > 0$ there exists an $N(\varepsilon)$ such that $\varepsilon > f_n(x) > 0$ for all $n \geq N(\varepsilon)$.

Since $E' = \bigcup_{j=1}^{\infty} V_j$ we conclude that if $x \in E'$ then $f_n(x)$ does not converge as $n \to \infty$ but that if $x \notin E'$ then $f_n(x) \to 0$. Thus

$$E' = \{x : f_n(x) \text{ diverges as } n \to \infty\}$$

which is the required result. □

The reader may find it helpful for following the rest of this essay to check how and at which point the argument above collapses if we omit any one of conditions (i), (ii), (iii) or (iv) from the conclusions of Lemma 2.4. In particular this may help her appreciate how closely the characterization of sets of convergence is connected with the study of bounded divergence.

§3 A Theorem of Zeller

Moving back to the special case of the convergence of power series we observe that (by replacing z by Rz if necessary) we may suppose that our power series has radius of convergence 1. Our problem may thus be considered one in Fourier analysis.

Problem. Study the convergence on $\mathbb{T} = [0,2\pi)$ of the partial sums $S_N = \sum_{r=0}^{N} a_r \exp ir\theta$. We adopt the usual convention of identifying θ with $\theta + 2n\pi$ $[n \in \mathbb{Z}]$ and putting the appropriate (quotient) topology on $T = \mathbb{R}/2\pi\mathbb{Z}$.

By exploiting a theorem of Kolmogorov, Zeller proved the following positive result.

Theorem 3.1'. If $E' \in F_\sigma$ then there exist $a_0, a_1, \ldots$ such that $E' = \{\theta : \sum_{r=0}^{N} a_r \exp i\theta$ converges$\}$. If fact the nature of the proof is such that it, almost automatically, produces a stronger result.

Theorem 3.1 (Zeller). If $E \in G_\delta$ then there exists an $f \in L^1(\mathbb{T})$ such that $\hat{f}(r) = 0$ for $r < 0$ and (writing $S_N(f,\theta) = \sum_{r=0}^{N} \hat{f}(r)\exp ir\theta$)

(i) $\limsup_{n\to\infty} |S_n(f,\theta)| = \infty$ for $\theta \in E$

(ii) $S_n(f,\theta)$ converges for $\theta \notin E$.

(Thus $\sum_{r=0}^{N} \hat{f}(r) \exp ir\theta$ diverges unboundedly on E and converges on E'.)

Proof of Theorem 3.1' from 3.1. Set $E = T \setminus E'$. □

The proof of Theorem 3.1 rests squarely on the following theorem which is due, to all intents and purposes, to Kolmogorov.

Theorem 3.2 (Kolmogorov). Given $[a,b] \subset \mathbb{T}$, $\varepsilon > 0$ and $K \geq 1$ we can find $a_0, a_1, \ldots, a_M \in \mathbb{C}$ such that writing $P(t) = \sum_{r=0}^{M} a_r \exp irt$ we have

(i) $\frac{1}{2\pi} \int_{\mathbb{T}} |P(t)|dt \leq \varepsilon$

(ii) $\max_{0 \leq m \leq M} \left| \sum_{r=0}^{m} a_r \exp irt \right| \leq \varepsilon$ for all $t \notin [a-\varepsilon, b+\varepsilon]$

(iii) $\max_{0 \leq m \leq M} \left| \sum_{r=0}^{m} a_r \exp irt \right| \geq K$ for all $t \in [a,b]$.

We defer the proof of Theorem 3.2 for the moment and instead show how the result can be used to prove Zeller's Theorem.

We need a simple topological lemma.

Lemma 3.3. (i) If U is an open subset of $\mathbb{T}$ then we can find a countable collection of disjoint open intervals $J_1, J_2, \ldots$ with $U = \bigcup_{k=1}^{\infty} J_k$.

(ii) If U is an open subset of T we can find closed intervals $[c_1,d_1], [c_2,d_2], \ldots$ together with real numbers $\varepsilon_1 > 0, \varepsilon_2 > 0, \ldots$ such that

(A) For each $j \geq 1$, $[c_j - \varepsilon_j, d_j + \varepsilon_j] \subseteq U$

(B) Each $t \in U$ belongs to at least one $[c_j, d_j]$

but

(C) Each $t \in U$ belongs to only finitely many $[c_j - \varepsilon_j, d_j + \varepsilon_j]$.

Proof. (i) This is standard. (For each rational $r \in E$ let $\alpha_r = \inf\{x < r : (x,r) \subseteq U\}$, $\beta_r = \sup\{r < y : (r,y) \subseteq U\}$. It is easy to check that $(\alpha_r, \beta_r) \subset U$, $\bigcup_{r \in E, r \in Q} (\alpha_r, \beta_r) = U$ and that if $r,s \in E \cap Q$ then either $(\alpha_r,\beta_r) = (\alpha_s,\beta_s)$ or $(\alpha_r,\beta_r)\ (\alpha_s,\beta_s) = \phi$. Since Q is countable the result follows.)

(ii) By (i) we need only consider the case when U is an open interval $J = (c,d)$ say. But it is then easy to write down sequences satisfying the given conditions. For example, taking $\eta = (d-c)$, we could set

$$[c_1,d_1] = [c+\delta/4, d-\delta/4], \qquad \varepsilon_1 = \delta/8$$

$$[c_{2r},d_{2r}] = [c+\delta/2^{r+1}, c+\delta/2^{r}], \qquad \varepsilon_{2r} = \delta/2^{4+2}$$

$$[c_{2r+1},d_{2r+1}] = [d-\delta/2^{r}, d-\delta/2^{r+1}], \qquad \varepsilon_{2r+1} = \delta/2^{r+2} \quad [r \geq 1]. \qquad \square$$

We can now set off on a familiar course.

Lemma 3.4. Given U an open subset of $\mathbb{T}$, $\varepsilon > 0$ and $K \geq 1$ we can find a sequence of trigonometric polynomials $P_j(t) = \sum_{r=o}^{M(j)} a_{jr} \exp irt$ such that

$$(i)_j \quad \frac{1}{2\pi}\int_{\mathbb{T}} |P_j(t)|dt \leq 2^{-j}\varepsilon,$$

$$(ii)_j \quad \max_{0 \leq m \leq M(j)} \left|\sum_{r=o}^{m} a_{jr} \exp irt\right| \leq 2^{-j}\varepsilon \text{ for all } t \notin U$$

(iii) For all $t \in U$ there exists at least one value of k such that $\max_{0 \leq m \leq M(k)} \left|\sum_{r=o}^{m} a_{jr} \exp irt\right| \geq K$,

(iv) For all $t \in U$ there exist only finitely many values of k such that $\max_{o \leq m \leq M(k)} \left|\sum_{r=o}^{m} \exp irt\right| > 2^{-k}\varepsilon$.

Proof. Choose $[c_j,d_j]$ and ε_j as in Lemma 3.3(ii). By Theorem 3.2 we can find $a_{jo}, a_{j1}, \dots, a_{j\,M(j)-1} \in \mathbb{C}$ such that, writing $P_j(t) = \sum_{r=o}^{M(j)} a_{jr} \exp irt$, we have

$$(i)_j' \quad \frac{1}{2\pi}\int_{\mathbb{T}} |P_j(t)|dt \leq 2^{-j}\varepsilon,$$

$$(ii)_j' \quad \max_{o \leq m \leq M(j)} \left|\sum_{r=o}^{m} a_{rj} \exp irt\right| \leq 2^{-j}\varepsilon$$

$$\text{for all } t \notin [c_j - \varepsilon_j, d_j + \varepsilon_j]$$

$(iii)_j{}'$ $\max_{0 \le m \le M(j)} \left| \sum_{r=0}^{m} a_r \exp irt \right| \ge K$ for all $t \in [c_j, d_j]$.

Condition $(i)_j$ is condition $(i)_j{}'$. Condition $(ii)_j$ follows from condition $(ii)_j{}'$ and the fact that $[c_j - \varepsilon_j, d_j + \varepsilon_j] \subseteq U$. Condition (iii) follows from condition $(iii)_k{}'$ and the fact that each $t \in U$ belongs to at least one $[c_k, d_k]$ whilst condition (iv) follows from condition $(ii)_k{}'$ and the fact that each $t \in U$ belongs to only finitely many $[c_k - \varepsilon_k, d_k + \varepsilon_k]$. □

<u>Lemma 3.5</u>. Given $U_1, U_2, \ldots$ open subsets of $\mathbb{T}$ we can find a sequence of trigonometric polynomials $Q_j(t) = \sum_{r=0}^{N(j)-1} b_{jr} \exp irt$ such that

(i) $\sum_{j=1}^{\infty} \frac{1}{2\pi} \int_{\mathbb{T}} |Q_j(t)|dt \le 1$

(ii) For all $t \in \bigcap_{n=1}^{\infty} U_n$

$$\sup_j \max_{0 \le m \le N(j)-1} \left| \sum_{r=0}^{m} b_{jr} \exp irt \right| = \infty$$

(iii) For all $t \notin \bigcap_{n=1}^{\infty} U_n$

$$\sum_{j=1}^{\infty} \max_{0 \le m \le N(j)-1} \left| \sum_{r=0}^{m} b_{jr} \exp irt \right| = \infty .$$

<u>Proof</u>. As in Lemma 2.5 we may assume $U_1 \supseteq U_2 \supseteq \ldots$. By Lemma 3.4 we can find trigonometric polynomials $P_{nj}(t) = \sum_{r=0}^{M(n,j)} a_{njr} \exp irt$ such that

$(i)_{nj}$ $\frac{1}{2\pi} \int_{\mathbb{T}} |P_{nj}(t)|dt \le 2^{-n-j}$,

$(ii)_{nj}$ $\max_{0 \le m \le M(n,j)} \left| \sum_{r=0}^{m} a_{njr} \exp irt \right| \le 2^{-n-j}$ for all $t \notin U_n$;

$(iii)_n$ For all $t \in U_n$ there exists at least one value of k such that $\max_{0 \le m \le M(n,k)} \left| \sum_{r=0}^{m} a_{nkr} \exp irt \right| \ge 2^n$,

$(iv)_n$ For all $t \in U_n$ there exist only finitely many values of k such that $\max_{0 \le m \le M(n,K)} \left| \sum_{r=0}^{m} a_{nkr} \exp irt \right| \ge 2^{-n-k}$.

From $(i)_{nj}$ we can conclude at once that

$$(i)' \quad \sum_{n=1}^{\infty} \sum_{j=1}^{\infty} \frac{1}{2\pi} \int_{\mathbb{T}} |P_{nj}(t)|\,dt \le 2^{-n-j}$$

If $t \in \bigcap_{n=1}^{\infty} U_n$ then, for each $n \ge 1$, it follows from $(iii)_r$ that we can find a $k(n)$ with

$$\max_{o \le m \le N(n,k(n))} \left| \sum_{r=o}^{m} a_{nk(n)r} \exp irt \right| \ge 2^n .$$

We conclude that

(ii)' If $t \in \bigcap_{n=1}^{\infty} U_n$ then

$$\sup_n \sup_k \max_{o \le m \le M(n,k)} \left| \sum_{r=o}^{m} a_{nkr} \exp irt \right| = \infty .$$

On the other hand if $t \notin \bigcap_{n=1}^{\infty} U_n$ then since $U_1 \supseteq U_2 \supseteq \dots$ we can find an N such that $t \notin U_n$ for all $n > N$. By $(ii)_{nj}$ and $(iv)_r$ we know that for each $1 \le n \le N$ we can find a $K(n)$ such that

$$\max_{o \le m \le M(n,k)} \left| \sum_{r=o}^{m} a_{nkr} \exp irt \right| \le 2^{-n-k} \quad \text{for } k > K(n)$$

whilst by $(iii)_n$ we know that for all $n > N$ and all $j \ge 1$

$$\max_{o \le m \le M(n,k)} \left| \sum_{r=o}^{m} a_{nkr} \exp irt \right| \le 2^{-n-k} .$$

Thus writing $K = \max_{1 \le k \le N} K(n)$ and writing

$$\Gamma = \{(n,k) : 1 \le k \le K,\ 1 \le n \le N\}$$

we have

$$\sum_{(n,k) \notin \Gamma} \max_{o \le m \le M(n,k)} \left| \sum_{r=o}^{m} a_{nkr} \exp irt \right| \le \sum_{(n,k) \notin \Gamma} 2^{-n-k} \le 1 .$$

Since Γ is finite it follows that

(iii)' For all $t \notin \bigcap_{n=1}^{\infty} U_n$ then

$$\sum_{n=1}^{\infty} \sum_{k=1}^{\infty} \max_{0 \le m \le M(n,k)} \left| \sum_{r=0}^{m} a_{nkr} \exp irt \right| < \infty .$$

We now take $Q_1, Q_2, \ldots$ to be some enumeration of the P_{nj}. Conditions (i), (ii) and (iii) can be read off directly from conditions (i)', (ii)' and (iii)'. □

<u>Proof of Theorem 3.1</u>. By definition $E = \bigcap_{n=1}^{\infty} U_n$ where $U_1, U_2, \ldots$ are open sets. Choose Q_j as in Lemma 3.5, write $M(k) = \sum_{j=1}^{k-1} N(j)$ and set $P_j(t) = \exp\big(iM(j)t\big)Q_j(t)$. Then

$$\sum_{j=1}^{\infty} \frac{1}{2\pi} \int_{\mathbf{T}} |P_j(t)|dt = \sum_{j=1}^{\infty} \frac{1}{2\pi} \int_{\mathbf{T}} |Q_j(t)|dt \le 1$$

and so $\sum_{j=1}^{\infty} P_j$ converges in L^1 to some function f with $\frac{1}{2n} \int |f(t)|dt \le 1$. By Lebesgue's theorem on dominated convergence $\hat{f}(r) = \lim_{N\to\infty} \sum_{j=1}^{N} \hat{P}_j(r) = \sum_{j=1}^{\infty} \hat{P}_j(r)$ and so $\hat{f}(r) = 0$ for $r \le 0$ whilst

$$\hat{f}\big(r+M(j)\big) = \hat{Q}_j(r) \quad \text{for all} \quad 0 \le r \le N(j) - 1.$$

It follows that, if $\theta \varepsilon \bigcap_{n=1}^{\infty} U_n$

$$\sup_j |S_{M(j)-1}(f,\theta) - S_{M(j+1)-1}(f,\theta)| = \sup_j \left| \sum_{r=0}^{N(j)-1} \hat{f}\big(M(j)+r\big) \exp\big(i(M(j)+r)\theta\big) \right|$$

$$= \sup_j \left| \sum_{r=0}^{N(j)-1} \hat{Q}_j(r) \exp\big(i(M(j)+r)\theta\big) \right|$$

$$= \sup_j \left| \sum_{r=0}^{N(j)-1} \hat{Q}_j(r) \exp ir\theta \right| = \infty$$

using condition (ii) of Lemma 3.5. Thus, if $\theta \varepsilon \bigcap_{n=1}^{\infty} U_n = E$, we see that $\sup_j |S_{M(j)-1}(f,\theta)| = \infty$.

Suppose on the other hand that $\theta \notin \bigcap_{n=1}^{\infty} U_n$. Then if $k \ge j$, $0 \le r \le N(j)-1$, $0 \le s \le N(k)-1$ (and, if $k = j$, $r \le s$) we have

$$|S_{M(k)+s}(f,\theta) - S_{M(j)+r}(f,\theta)|$$

$$\leq |S_{M(j)+r}(f,\theta)-S_{M(j)-1}(f,\theta)| + \sum_{\ell=j}^{k-1} |S_{M(\ell+1)-1}(f,\theta)-S_{M((\ell)-1}(f,\theta)| +$$

$$+ |S_{M(k)+s}(f,\theta) - S_{M(k)-1}(f,\theta)|$$

$$= |\sum_{u=0}^{r} \hat{Q}_j(u)\exp iu\theta| + \sum_{\ell=j}^{k-1} |\sum_{u=0}^{N(\ell)-1} \hat{Q}_\ell(u)\exp iu\theta| + |\sum_{u=0}^{s} \hat{Q}_k(u)\exp iu\theta|$$

$$\leq 3 \sum_{\ell=j}^{\infty} \max_{0\leq m\leq N(\ell)-1} |\sum_{u=0}^{m} b_{\ell u} \exp iu\theta| \to 0$$

as $j \to \infty$ by condition (iii) of Lemma 3.5. Thus by the general principle of convergence $S_n(f,\theta)$ converges for all $\theta \notin E$. □

In conclusion we remark that if we were interested in unbounded divergence rather than divergence, Zeller's theorem would show that only topological restrictions entered into the problem.

Lemma 3.6. If (X,d) is a metric space and $f_1, f_2, \ldots$ are continuous maps from X to $\mathbb{C}$ then $\{x : \sup_n |f_n(x)| = \infty\} \in G_\delta$.

Proof. Observe that

$$\{x : \sup_n |f_n(x)| = \infty\} = \bigcap_{m=1}^{\infty} \bigcup_{n=1}^{\infty} \{x : |f_n(x)| > m\}$$

$$\in G_{\sigma\delta} = G_\delta$$ □

§4 Kolmogorov's Theorem

We still have to prove Kolmogorov's Theorem (Theorem 3.2). This we do by making use of a theorem of Kronecker of considerable importance both in number theory and harmonic analysis.

<u>Definition 4.1</u>. Points $x_1, x_2, \ldots, x_n \in \mathbb{T}$ are said to be (linearly) independent (over Q) if the only integer solution of

$$m_1x_1 + m_2x_2 + \ldots + m_nx_n = 0$$

is the trivial one $m_1 = m_2 = \ldots = m_n = 0$.
(Thus, for example $x_1 = \pi + \sqrt{2} + 1$, $x_2 = \sqrt{2} - 1$, $x_3 = 3/2$ do not form an independent set since $6x_1 + (1-6)x_2 + 4x_3 = 0$. On the other hand each of the pairs x_1, x_2; x_2, x_3 and x_3, x_1 are independent.)

<u>Theorem 4.2 (Kronecker)</u>. If $x_1, x_2, \ldots, x_n$ are independent then given $\theta_1, \theta_2, \ldots, \theta_n \in \mathrm{T}$ and $\varepsilon > 0$ we can find an $N \geq 0$ such that

$$|N_jx_j - \theta_j| < \varepsilon \qquad \text{for all} \quad 1 \leq j \leq n .$$

In other words if I start n clocks all running at different speeds (with the minute hand of the jth hand covering x_j minute on the clock face each hour say), then unless some special relation exists between the speeds (i.e. unless the x_j are linearly independent over Q) then after some number N of hours have passed the minute hands will point in some previously specified direction (θ_j minute past the hour, say) to within any desired degree of accuracy.

There are many interesting proofs of Kronecker's theorem. We shall prove it via a celebrated extension due to Weyl.

<u>Theorem 4.3 (Weyl)</u>. If $x_1, x_2, \ldots, x_n$ are independent and $f : \mathbb{T}^n \to \mathbb{C}$ is continuous then

$$\frac{1}{M}\sum_{r=0}^{M-1} f(rx_1, rx_2, \ldots, rx_n) \to \left(\frac{1}{2\pi}\right)^n \int_{\mathbb{T}^n} f(t_1, t_2, \ldots, t_n)\,dt_1\,dt_2 \ldots dt_n$$

as $M \to \infty$.

<u>Proof of Theorem 4.2 from Theorem 4.3.</u> Choose a continuous function $f : \mathbb{T}^n \to \mathbb{R}$ in such a way that

(i) $1 \geq f(t_1, t_2, \ldots, t_n) \geq 0$ for all $t_1, t_2, \ldots, t_n$

(ii) $f(t_1, t_2, \ldots, t_n) \neq 0$ only if $|t_j - \theta_j| < \varepsilon$ for each $1 \leq j \leq n$

(iii) $\left(\frac{1}{2\pi}\right)^n \int_{\mathbb{T}} f(t_1, t_2, \ldots, t_n)\, dt_1\, dt_2 \ldots dt_n = 1$.

Then $\frac{1}{M} \sum_{r=o}^{M-1} f(rx_1, rx_2, \ldots, rx_n) \to 1$ so there must exist at least one $N \geq 0$ with $f(Nx_1, Nx_2, \ldots, Nx_n) \neq 0$. Using (ii) we see that $|Nx_j - \theta_j| < \varepsilon$ for each $1 \leq j \leq n$. □

(A closer look shows that Weyl's theorem is an ergodic theorem which says that not only are the $(rx_1, rx_2, \ldots, rx_n)$ dense in $\mathbb{T}^n$ but that, as r changes, they appear in each well behaved region the "correct" proportion of the time.)

<u>Proof of Theorem 4.3.</u> Set

$$T_M f = \frac{1}{M} \sum_{r=o}^{M-1} f(rx_1, rx_2, \ldots, rx_n) - \left(\frac{1}{2\pi}\right)^m \int_{\mathbb{T}^n} f(t_1, t_2, \ldots, t_n)\, dt_1 dt_2 \ldots dt_n .$$

Then $T_M 1 = 0$ and if e is an exponential function with $e(t_1, t_2, \ldots, t_n) = \exp i\,(m_1 t_1 + m_2 t_2 + \ldots + m_n t_n)$ say $[n_j \in \mathbb{Z},$ not all the n_j zero$]$ we have

$$T_M e = \frac{1}{M} \sum_{r=o}^{M-1} \left(\exp i\,(m_1 x_1 + m_2 x_2 + \ldots + m_n x_n)\right)^r - 0$$

$$= \frac{1}{M} \frac{1 - \exp i\, M(m_1 x_1 + m_2 x_2 + \ldots + m_n x_n)}{1 - \exp i\,(m_1 x_1 + m_2 x_2 + \ldots + m_n x_n)}$$

$\to 0$ as $M \to \infty$. (Observe that the condition $m_1 x_1 + m_2 x_2 + \ldots + m_n x_n \neq 0$ is essential for the argument.)

Since T_M is linear it follows that when P is any trigonometric polynomial (i.e. sum of scalar multiples of exponentials) $T_M P \to 0$. But the trigonometric polynomials are uniformly dense in the continuous functions (by the theorem of Stone Weiersbrass or otherwise) and $|T_M f| \leq 2\|f\|_\infty$ (where as usual $\|f\|_\infty = \sup|f(t_1,t_2,\ldots,t_n)|$) so $T_M f \to 0$ for all continuous f. □

We shall also need two simple lemmas concerned with independence.

Lemma 4.4. Given $x_1,x_2,\ldots,x_{n-1}$ independent points of $\mathbf{T}$, $y \in \mathbf{T}$ and $\eta > 0$ we can find x_n with $|x_n - y| < \eta$ and $x_1,x_2,\ldots,x_n$ independent.

Proof. There are a countable number of sets

$$E(m_1,m_2,\ldots,m_n) = \{W : m_1x_1 + m_2x_2 + \ldots + m_{n-1}x_{n-1} + m_nW = 0\}$$

with $m_1,m_2,\ldots,m_n \in \mathbf{Z}$ and $m_n \neq 0$. Each $E(m_1,m_2,\ldots,m_n)$ is finite so taking E to be the union of these $E(m_1,m_2,\ldots,m_n)$ we know that E is countable. Thus $(y-\eta, y+\eta) \cap E \neq \emptyset$ and choosing $x_n \in (y-\eta, y+\eta) \cap E$ we have a point with the desired properties. □

The translate of an independent set is not necessarily independent, but the next lemma shows that it is "almost independent" and that essentially only one linear relation (with integer coefficients) can exist for the translated set.

Lemma 4.5. Suppose $x_1,x_2,\ldots,x_n$ are independent and $t \in \mathbf{T}$. Suppose further that $m_1,m_2,\ldots,m_n \in \mathbf{Z}$ and are not all zero and $m_1',m_2',\ldots,m_n' \in \mathbf{Z}$ and are not all zero yet

$$m_1(x_1 - t) + m_2(x_2 - t) + \ldots + m_n(x_n - t) = 0 \qquad (*)$$

$$m_1'(x_1-t) + m_2'(x_2-t) + \ldots + m_n(x_n - t) = 0 \qquad (*)'$$

Then there exist non zero integers M and M' with $Mm_j' = M'm_j$ for each $1 \leq j \leq n$.

Proof. Let $M = m_1 + m_2 + \ldots + m_n$. Then if $M = 0$, equation (*) reduces to $m_1x_1 + m_2x_2 + \ldots + m_nx_n = 0$ which is excluded by our hypotheses. Thus $M \neq 0$ and similarly $M' = m_1' + m_2' + \ldots + m_n' \neq 0$. By (*) and (**)'

$$\sum_{j=1}^{n} M'm_jx_j = M'Mt = \sum_{j=1}^{n} M m_j' x_j$$

so that $\sum_{j=1}^{n} (M'm_j - Mm_j')\, x_j = 0$ and by the independence of $x_1, x_2, \ldots, x_n$ we have $M'm_j - Mm_j' = 0$ for all $1 \leq j \leq n$ as required. □

The next lemma is the key step in the proof of Kolmogorov's theorem. (Here δ_x is the Dirac point mass at x.)

Lemma 4.6. Given $[a,b] \subseteq \mathbb{T}$ and $K \geq 1$ we can find an integer $n \geq 1$ and points $x(1), x(2), \ldots, x(n) \in (a,b)$ such that if $\mu = n^{-1} \sum_{j=1}^{n} \delta_{x(j)}$, we have

$$\sup_{N \geq 0} \left| \sum_{r=-N}^{N} \hat{\mu}(r) \exp irt \right| \geq 4K \qquad \text{for each } t \in [a,b] .$$

Proof. Fix $n \geq 100$ for the time being and write $\eta = b - a$. By repeated application of Lemma 4.4 we can find $x(1), x(2), \ldots, x(n)$ independent such that

$$\left| x(j) - \left(a + \frac{(2j-1)}{2n+2}\eta \right) \right| < \frac{\eta}{100} \qquad [1 \leq j \leq n] . \tag{*}$$

We observe that

$$\hat{\mu}(r) = n^{-1} \sum_{j=1}^{n} \hat{\delta}_{x(j)}(r) = n^{-1} \sum_{j=1}^{n} \exp(-ir\, x(j))$$

and so, writing $S_N(\mu,t) = \sum_{r=-N}^{N} \hat{\mu}(r)\exp irt$, we have

$$S_N(\mu,t) = n^{-1} \sum_{j=1}^{n} \sum_{r=-N}^{N} \exp\left(-ir(x(j)-t)\right)$$

$$= n^{-1} D_N\left(x(j)-t\right)$$

where $D_N\left(x(j)-t\right) = \dfrac{\sin(N+\frac{1}{2})\left(x(j)-t\right)}{\sin\frac{1}{2}\left(x(j)-t\right)}$ for $x(j)-t \neq 0$ and $D_N(0) = 2N+1$.

By renumbering, if necessary, we may assume that $|x(1)-t| \leq |x(2)-t| \leq \ldots \leq |x(n)-t|$ if $t \in [a,b]$ condition (*) shows that (after renumbering) $|x(j)-t| \leq j\eta/n$ and that $|x(2)-t| \geq \eta/4n$ The inequality $|s| \geq \sin|s| \geq 4|s|/\pi$ for $|s| \leq \pi/2$ now shows that

$$|\sin\tfrac{1}{2}\left(x(j)-t\right)|^{-1} \geq n\eta^{-1}/j \qquad [1 \leq j \leq n]$$

but

$$|\sin\tfrac{1}{2}\left(x(k)-t\right)|^{-1} \leq 8n\eta^{-1} \qquad [2 \leq k \leq n] .$$

We want to show that if $t \in [a,b]$ then $S_N(\mu,t)$ is large for some N. We shall consider 3 cases of which the first is both typical and simple.

<u>Case A.</u> Suppose $x(1)-t, x(2)-t, \ldots, x(n)-t$ are independent. Then by Kronecker's theorem (Theorem 4.2) we can find an $N \geq 0$ such that

$$|N\left(x(j)-t\right) - \left(\frac{\pi}{2} - \left(x(j)-t\right)/2\right)| \leq 10^{-1}$$

so that

$$\sin\ (N+\tfrac{1}{2})\left(x(j)-t\right) \geq 1/2$$

and

$$D_N\left(x(j)-t\right) \geq |\left(\sin\tfrac{1}{2}\left(x(j)-t\right)\right)|^{-1}/2$$

for all $1 \leq j \leq n$. Thus using the results of the last paragraph

$$D_N\left(x(j)-t\right) \geq n\eta^{-1}/2j \qquad [1 \leq j \leq n]$$

and

$$S_N(\mu,t) \geq n^{-1} \sum_{j=1}^{n} n\eta^{-1}/2j = \eta^{-1}2^{-1} \sum_{j=1}^{n} n^{-1} \geq (2\eta)^{-1} \log n .$$

Case B. Suppose that $\sum_{j=1}^{n} m(j)(x(j)-t) = 0$ for some integers $m(1), m(2), \ldots, m(n)$ where $m(k) \neq 0$ for some $k \neq 1$. By Lemma 4.5 the points $x(1)-t, x(2)-t, \ldots, x(k-1)-t, x(k+1)-t, \ldots, x(n)-t$ must be independent and so arguing as in Case A we can find an $N \geq 0$ such that

$$D_N(x(j)-t) \geq n\eta^{-1}/2j \qquad [j \neq k].$$

On the other hand since $k \geq 2$ we know that

$$D_N(x(k)-t) \geq -|\sin \tfrac{1}{2}(x(k)-t)| \geq -8n\eta^{-1} \geq -9n\eta^{-1} + n\eta^{-1}/2k$$

so $$S_N(\mu,t) \geq n^{-1}\left(\sum_{j=1}^{n} n\eta^{-1}/2j - 9n\eta^{-1}\right) \geq (2\eta)^{-1} \log n - 9\eta^{-1} .$$

Case C. The only remaining possibility is that $m(1)(x(1)-t) = 0$ for some integer $m(1) \neq 0$. Replacing $m(1)$ by $-m(1)$ if necessary we may suppose $m(1) > 0$. By Lemma 4.2 the points $x(2)-t, x(3)-t, \ldots, x(n)-t$ must be independent and so, fairly obviously, the points $m(1)(x(2)-t), m(1)(x(3)-t), \ldots, m(1)(x(n)-t)$ must be independent. By Kronecker's theorem we can find an $N' \geq 1$ such that

$$|N'm(1)(x(j)-t) - (\tfrac{\pi}{2} - (x(j)-t)/2)| < 1/10 \qquad [2 \leq j \leq n]$$

and so writing $N = N'm(1)$ we have

$$D_N(x(j)-t) \geq n\eta^{-1}/2j \qquad [2 \leq j \leq n] .$$

On the other hand $N(x(1)-t) = N'(m(1)(x(1)-t)) = 0$ and $|(x(1)-t)/2| \leq \eta/2n \leq 1/10$ so

$$|N'm(1)(x(1)-t) - (-(x(1)-t)/2)| < 1/10,$$

$$\sin\left((N+\tfrac{1}{2})(x(1)-t)\right) \geq 0$$

and
$$D_N(x(1)-t) \geq 0 = \eta^{-1}/2 - \eta^{-1}/2$$

Thus $S_N(\mu,t) \geq (2\eta)^{-1}\log n - \eta^{-1}/2$.

Summary. We have seen that, for each $t \in [a,b]$,

$$\sup_{N \geq 0} |S_N(\mu,t)| \geq \eta^{-1}(2^{-1}\log n - 9).$$

Taking n large enough so that $2^{-1}\log n - 9 \geq 4K$ gives the required result. □

The remainder of the proof of Kolmogorov's theorem is just tinkering.

Lemma 4.7. Given $[a,b] \subseteq \mathbb{T}$ and $K \geq 1$ we can find an integer $n \geq 1$, points $x(1), x(2), \ldots, x(n) \in (a,b)$ and an integer $M \geq 1$ such that if $\mu = n^{-1}\sum_{j=1}^{n}\delta_{x(j)}$, we have

$$\max_{m \geq N \geq 0} \left|\sum_{r=-N}^{N}\hat{\mu}(r)\exp irt\right| \geq 2K \quad \text{for each } t \in [a,b].$$

Proof. Choose n and $x(1), x(2), \ldots, x(n)$ as in Lemma 4.7. For each $t \in [a,b]$ we can find an $N(t) \geq 0$ such that $|S_{N(t)}(\mu,t)| \geq 3K$. Since $S_{N(t)}(\mu, \ )$ is continuous we can find an $\varepsilon(t) > 0$ such that $|S_{N(t)}(\mu,x)| \geq 2K$ for all $|x-t| < \varepsilon(t)$. The intervals $I_t = (t-\varepsilon(t), t+\varepsilon(t))$ form an open cover of the compact set $[a,b]$ and so we can find a finite sub cover $I_{t(1)}, I_{t(2)}, \ldots, I_{t(q)}$ say. Set $M = \max_{1 \leq p \leq q} N(t(p))$. □

Lemma 4.8. Given $[a,b] \subseteq \mathbf{T}$ and $K \geq 1$ we can find a twice continuously differentiable function f and an integer $M \geq 1$ such that

(i) $\frac{1}{2\pi} \int_{\mathbf{T}} |f(t)|dt = 1$

(ii) $f(t) = 0$ for $t \notin [a,b]$

(iii) $\max_{M \geq N \geq 0} |\sum_{r=-N}^{N} \hat{f}(r) \exp irt| \geq K$ for $t \in [a,b]$.

Proof. Choose a sequence of twice continuously differentiable functions h_m such that

(1) $h_m(t) \geq 0$ for all $t \in \mathbf{T}$

(2) $h_m(t) = 0$ for $|t| \geq 1/m$

(3) $\frac{1}{2\pi} \int_{\mathbf{T}} h_m(t)dt = 1.$

Let μ and M be chosen as in Lemma 4.6 and set $f_m = \mu * h_m$. Automatically $f_m(t) \geq 0$ for all $t \in \mathbf{T}$ and $\hat{f}_m(r) = \hat{\mu}(r)\hat{h}_m(r)$. Since

$$\frac{1}{2\pi} \int_{\mathbf{T}} |f_m(t)|dt = \frac{1}{2\pi} \int_{\mathbf{T}} f_m(t)dt = \hat{f}_m(0) = \hat{\mu}(0)\hat{h}_m(0) = 1$$

we have

(i)' $\frac{1}{2\pi} \int_{\mathbf{T}} |f_m(t)|dt = 1.$

Trivially

(ii)' $f_m(t) = 0$ if $t \notin \bigcup_{j=1}^{n} (x(j) - 1/m,\ x(j) + 1/m)$.

Finally (by observing that $h_m \to \delta_o$ weakly or by simple direct calculation) we see that $\hat{h}_m(r) \to 1$ so $\hat{f}_m(r) \to \hat{\mu}(r)$ and

$$\text{(iii)}' \quad \max_{M \geq N \geq 0} \left| \sum_{r=-N}^{N} \hat{f}_m(r) \exp irt \right| \geq \max_{M \geq N \geq 0} \left| \sum_{r=-N}^{N} \hat{\mu}(r) \exp irt \right| - \sum_{r=-M}^{M} |\hat{f}_m(r) - \hat{\mu}(r)|$$

$$\geq 2K - \sum_{r=-M}^{M} |\hat{f}_m(r) - \hat{\mu}(r)| \to 2K$$

as $m \to \infty$.

Thus taking m sufficiently large and writing $f = f_m$, conditions (i), (ii) and (iii) may be read off directly from conditions (i)', (ii)' and (iii)'. □

<u>Lemma 4.9</u>. Suppose $[a,b] \subset \mathbb{T}$, $\varepsilon > 0$ and $f : \mathbb{T} \to \mathbb{C}$ is a continuous function such that

(i) $\frac{1}{2\pi} \int_{\mathbb{T}} |f(t)| dt = 1$

and

(ii) $f(t) = 0$ for $t \notin [a,b]$.

Then $\left| \sum_{r=N(1)}^{N(2)} \hat{f}(r) \exp irt \right| \leq 4\varepsilon^{-1}$ for all $t \notin [a-\varepsilon, b+\varepsilon]$ and all $N(2) > N(1)$.

<u>Proof</u>. Set $N = N(2) - N(1) - 1$. We observe that

$$\left| \sum_{r=N(1)}^{N(2)} \hat{f}(r) \exp irt \right| = \frac{1}{2\pi} \int_{\mathbb{T}} f(x) \sum_{r=N(1)}^{N(2)} \exp ir(t-x) dx \Big|$$

$$\leq \frac{1}{2\pi} \int_{\mathbb{T}} |f(x)| \left| \sum_{r=N(1)}^{N(2)} \exp ir(t-x) \right| dx$$

$$= \frac{1}{2\pi} \int_a^b |f(x)| \left| \frac{\sin(n+\frac{1}{2})(t-x)}{\sin\frac{1}{2}(t-x)} \right| dx$$

$$\leq \frac{1}{2} \int_a^b |f(x)| dx \sup_{x \in [a,b]} \left| \frac{\sin(N+\frac{1}{2})(t-x)}{\sin\frac{1}{2}(t-x)} \right|$$

$$\leq \frac{1}{\sin \varepsilon/2} \leq 4\varepsilon^{-1}, \quad \text{as stated.}$$ □

<u>Lemma 4.10</u>. Given $[a,b] \subset \mathbb{T}$, $1 > \varepsilon > 0$ and $K \geq 1$ we can find a twice continuously differentiable function g and an integer $L \geq 1$ such

that

(i) $\frac{1}{2\pi}\int_{\mathbf{T}} |g(t)|dt \leq \varepsilon/2$

(ii) $\max_{N(2)\geq N(1)} \left|\sum_{r=N(1)}^{N(2)} \hat{g}(r)\exp irt\right| \leq \varepsilon/2$ for all $t \notin [a-\varepsilon, b+\varepsilon]$

(iii) $\max_{L\geq N\geq 0} \left|\sum_{r=-N}^{N} \hat{g}(r)\exp irt\right| \geq 2K+\varepsilon/2$ for all $t \in [a,b]$.

<u>Proof</u>. By Lemma 4.7 we can find a twice continuously differentiable function f and an integer $L \geq 1$ such that

(i)' $\frac{1}{2\pi}\int_{\mathbf{T}} |f(t)|dt = 1$

(ii)' $f(t) = 0$ for $t \notin [a,b]$

(iii)' $\max_{L\geq N\geq 0} \left|\sum_{r=-N}^{N} \hat{f}(r)\exp irt\right| \geq 8\varepsilon^{-2}(2K+\varepsilon/2)$.

Using (i)', (ii)" and Lemma 4.8 we obtain

(ii)' $\left|\sum_{r=N(1)}^{N(2)} \hat{f}(r)\exp irt\right| \leq 4\varepsilon^{-1}$ for all $t \in [a-\varepsilon, b+\varepsilon]$

and all $N(2) > N(1)$.

Setting $g = 8^{-1}\varepsilon^{2}f$ we can read off conditions (i), (ii) and (iii) from conditions (i)', (ii)' and (iii)'. (Note that if $N(1) = N(2)$ $\left|\sum_{r=N(1)}^{N(2)} \hat{g}(r)\exp irt\right| = |\hat{g}(N(1))| \leq \frac{1}{2\pi}\int |g(t)|dt \leq \varepsilon/2$.) □

<u>Proof of Theorem 3.2</u>. Choose g as in Lemma 4.10. Since g is twice continuously differentiable $|\hat{g}(r)| \leq Ar^{-2}$ for some constant A and all $r \neq 0$. We can thus find an integer $L(0) \geq L$ such that $\sum_{|r|>L(0)} |\hat{g}(r)| \leq \varepsilon/2$. Setting $Q(t) = \sum_{r=-L(0)}^{L(0)} \hat{g}(r)\exp irt$ conditions (i), (ii) and (iii) of Lemma 4.9 give, immediately,

(i)' $\frac{1}{2\pi}\int_{\mathbf{T}} |Q(t)|dt \leq \varepsilon$

(ii)' $\max_{N(2)\geq N(1)} \left|\sum_{r=N(1)}^{N(2)} \hat{Q}(r)\exp irt\right| \leq \varepsilon$ for all $t \notin [a-\varepsilon, b+\varepsilon]$

(iii)' $\max_{L(0)\geq N\geq 0} \left|\sum_{r=-N}^{N} \hat{Q}(r)\exp irt\right| \geq 2K$ for all $t \in [a,b]$.

(To see (i)' observe that $\left|\sum_{|r|>L(0)} \hat{g}(r)\exp irt\right| \leq \varepsilon/2$ for $t \in \mathbf{T}$.)

Now set $P(t) = (\exp iL(0)t)Q(t)$ and $M = 2L(0)+1$. Automatically $P(t) = \sum_{r=o}^{M} a_r \exp irt$ for suitable $a_r \in \mathbb{C}$, whilst

(i) $\frac{1}{2\pi}\int_{\mathbf{T}} |P(t)|dt = \frac{1}{2\pi}\int_{\mathbf{T}} |Q(t)|dt \leq \varepsilon$,

(ii) $\max_{o\leq m\leq M} \left|\sum_{r=o}^{m} a_r \exp irt\right| = \max_{-L(0)\leq n\leq L(0)} \left|\sum_{r=-L(0)}^{n} \hat{Q}(r)\exp irt\right| \leq \varepsilon$

and

(iii)'' $\max_{L(0)\geq N\geq 0} \left|\sum_{r=o}^{N-1} a_r \exp irt - \sum_{r=o}^{2L(0)-N} a_r \exp irt\right|$

$$= \max_{L(0)\geq N\geq 0} \left|\sum_{r=N-L(0)}^{L(0)-N} \hat{Q}(r)\exp irt\right| \geq 2K$$

so that

(iii) $\max_{M\geq N\geq 0} \left|\sum_{r=o}^{N} a_r \exp irt\right| \geq K$ for all $t \in [a,b]$.

The proof is complete. □

§§5. Theorem of Marcinkiewicz

So far we have dealt with what a trigonometric series or power series can do. We now consider what they cannot do. The proof of the following theorem was outlined by Marcinkiewicz (though as will soon appear one essential fact was still unproved when he wrote.)

Theorem 5.1 (Marcinkiewicz). If $a_n \to 0$ as $|n| \to \infty$ (so in particular if $a_n = \hat{f}(n)$ with $f \in L^1$) then $\sum_{-N}^{N} a_n \exp int$ cannot diverge boundedly everywhere as $N \to \infty$.

The proof requires 3 preliminary results each of them interesting in their own right.

Lemma 5.2. If $\sum_{-N}^{N} a_n \exp int$ is bounded at each t independent of N (i.e. if we can find a $K(t)$ with $|\sum_{-N}^{N} a_n \exp int| \leq K(t)$ for all $N \geq 0$) then there is an interval $J = (a,b)$ on which $\sum_{-N}^{N} a_n \exp int$ is uniformly bounded (i.e. there is a K such that $|\sum_{-N}^{N} a_n \exp int| \leq K$ for all $t \in J$ and $N \geq 0$).

Proof. Let $f_N(t) = \sum_{-N}^{N} a_n \exp int$. Since f_N is continuous $E_{NM} = \{t : |f_N(t)| \leq M\}$ is closed and so

$$E_M = \bigcap_{N=1}^{\infty} \{t : |f_N(t)| \leq M\} = \{t : |f_N(t)| \leq M \text{ for all } M\} \text{ is closed.}$$

But $\bigcup_{M=1}^{\infty} E_M = \mathbf{T}$ so by the Baire Category theorem there must exist a K for which E_K contains an interval J. □

The next result is the Riemann localisation lemma in a form known to Riemann but stronger than that given in some elementary texts If $a_n \to 0$ as $|n| \to \infty$ and $\phi : \mathbf{T} \to \mathbf{C}$, is three times continuously differentiable, let us write, purely formally, $\hat{S}(n) = a_n$ and

$$\widehat{\phi S}(n) = \sum_{m=-\infty}^{\infty} \hat{\phi}(n-m)\hat{S}(m) \quad (\text{since } \hat{\phi}(r) = O(|r|^{-3}) \text{ the sum converges}).$$

Today we would call S a distribution, but, though this suggests why Theorem 5.3 might be true, the actual proof makes use of nothing beyond the formal definitions.

Theorem 5.3 (Riemann). If $a_n \to 0$ as $|n| \to \infty$ and $\phi : \mathbf{T} \to \mathbf{C}$ is three times continuously differentiable then

$$\left| \sum_{n=-N}^{N} \widehat{\phi S}(n) \exp int - \phi(t) \sum_{n=-N}^{N} \hat{S}(n) \exp int \right| \to 0$$

uniformly as $N \to \infty$.

Corollary 5.4. If $a_n \to 0$ as $|n| \to \infty$ and $\sum_{n=-N}^{N} a_n \exp int$ is uniformly bounded on $J = (a,b)$ then we can find $b_n \in \mathbf{C}$ such that

(i) $\sum_{n=-N}^{N} b_n \exp int$ is uniformly bounded on $\mathbf{T}$

(ii) $\sum_{n=-N}^{N} b_n \exp int \to 0$ uniformly an $N \to \infty$ for all $t \notin J$

(iii) If $t \in J$ then $\sum_{n=-N}^{N} a_n \exp int$ and $\sum_{n=-N}^{N} b_n$ converge and diverge together.

Proof of Corollary 5.4. Choose $\phi : \mathbf{T} \to \mathbf{R}$ a three times continuously differentiable function such that $1 \geq \phi(t) \geq 0$ for all $t \in \mathbf{T}$, $\phi(t) = 0$ for $t \notin J$ and $\phi(t) \neq 0$ for $t \in J$. Set $b_n = \widehat{\phi S}(n)$. Since

$$\left| \sum_{n=-N}^{N} b_n \exp int - \phi(t) \sum_{n=-N}^{N} a_n \exp int \right| \to 0$$ uniformly the results can be read off directly. □

Proof of Theorem 5.3. This is a matter of careful computation. Since $|\hat{\phi}(r)| \leq A|r|^{-3}$ $[r \neq 0]$ for some constant A all the series involved are absolutely convergent and this will justify the various changes of order of summation and so on. Note first that since $\sum_{n=-\infty}^{\infty} |\hat{\phi}(n)|$ converges

$$\phi(t) = \sum_{n=-\infty}^{\infty} \hat{\phi}(m) \exp imt .$$

Thus

$$\left| \sum_{n=-N}^{N} \widehat{\phi S}(n) \exp int - \phi(t) \sum_{n=-N}^{N} \hat{S}(n) \exp int \right|$$

$$= \left| \sum_{n=-N}^{N} \sum_{r=-\infty}^{\infty} \hat{\phi}(n-r)\hat{S}(r) \exp int - \sum_{n=-\infty}^{N} \sum_{n=-N}^{N} \hat{\phi}(m)\hat{S}(n) \exp i(n+m)t \right|$$

$$= \left| \sum \sum_{|n+m| \leq N} \hat{\phi}(m)\hat{S}(n) \exp i(n+m)t - \sum \sum_{|n| \leq N} \hat{\phi}(m)\hat{S}(n) \exp i(n+m)t \right|$$

$$= \left| \sum_{(n,m) \in \Gamma} \hat{\phi}(m)\hat{S}(n) \exp i(n+m)t \right| \leq \sum_{(n,m) \in \Gamma} |\hat{\phi}(m)\hat{S}(n)|$$

$$= \sum_{j=1}^{4} \sum_{(n,m) \in \Gamma(j)} |\hat{\phi}(m)\hat{S}(n)|$$

where $\Gamma = \bigcup_{j=1}^{4} \Gamma(j)$ and

$$\Gamma(1) = \{(n,m) : n+m > N,\ N \geq n \geq -N\}$$

$$\Gamma(2) = \{(n,m) : n+m < -N,\ N \geq n \geq -N\}$$

$$\Gamma(3) = \{(n,m) : -N \leq n+m \leq N,\ m > 0\}$$

$$\Gamma(4) = \{(n,m) : -N \leq n+m \leq N,\ m < 0\}.$$

Now $\Gamma(3) \subseteq \{(n,m) : m \geq -N+n,\ n \geq N+1\}$ so

$$\sum_{\Gamma(3)} |\hat{\phi}(m)\ \hat{S}(n)| \leq \sum_{n=N+1}^{\infty} \sum_{m=-N+n}^{\infty} |\hat{\phi}(m)|\,|\hat{S}(n)|$$

$$\leq \sup_{|r| \geq N+1} |\hat{S}(r)| \sum_{n=N+1}^{\infty} \sum_{m=-N+n}^{\infty} A\,|m|^{-3}$$

$$\leq \sup_{|r| \geq N+1} |\hat{S}(r)| \sum_{p=1}^{\infty} \sum_{m=p}^{\infty} A\, m^{-3}$$

$$\leq \sup_{|r| \geq N+1} |\hat{S}(r)| \sum_{p=1}^{\infty} A'\, p^{-2}$$

$$\leq A'' \sup_{|r| \geq N+1} |\hat{S}(r)|$$

where A' and A'' are constants. Similarly

$$\sum_{\Gamma(4)} |\hat{\phi}(m)\ \hat{S}(n)| \leq A'' \sup_{|r| \geq N+1} |\hat{S}(r)|\,.$$

Our estimates in the remaining two cases are slightly more complex. Choose some M with $N \geq M \geq 1$.

$$\sum_{\Gamma(1)} |\hat{\phi}(m)\hat{S}(n)| = \sum_{N \geq |n| \geq M} \sum_{n+M>N} |\hat{\phi}(m)|\,|\hat{S}(n)| + \sum_{n=-M+1}^{M-1} \sum_{n+m>N} |\hat{\phi}(m)|\,|\hat{S}(n)|$$

$$\leq \sup_{|r| \geq M} |\hat{S}(r)| \sum_{N \geq |n| \geq M} \sum_{n+m>N} A|m|^{-3} + \sup_{r \in \mathbb{Z}} |\hat{S}(r)| \sum_{n=-M+1}^{M-1} \sum_{n+m>N} A|m|^{-3}$$

$$\leq \sup_{|r|\geq M} |\hat{S}(r)| \sum_{n=-\infty}^{N} \sum_{n+m>N} A|m|^{-3} + \sup_{r\varepsilon \mathbb{Z}} |\hat{S}(r)| \sum_{n=-\infty}^{M-1} \sum_{n+m>N} A|m|^{-3}$$

$$= \sup_{|r|\geq M} |\hat{S}(r)| \sum_{p=1}^{\infty} \sum_{m=p}^{\infty} A\, m^{-3} + \sup_{r\varepsilon Z} |\hat{S}(r)| \sum_{p=N-M+1}^{\infty} \sum_{m=p}^{\infty} A\, m^{-3}$$

$$\leq \sup_{|r|\geq M} |\hat{S}(r)| \sum_{p=1}^{\infty} A' p^{-2} + \sup_{r\varepsilon Z} |\hat{S}(r)| \sum_{p=N-M+1}^{\infty} A' p^{-2}$$

$$\leq A'' \sup_{|r|\geq M} |S(r)| + A''' \sup_{r\varepsilon Z} |\hat{S}(r)| (N-M+1)^{-1} .$$

where A''' is a constant. Similarly

$$\sum_{\Gamma(2)} |\hat{\phi}(m)\hat{S}(n)| \leq A'' \sup_{|r|\geq M} |\hat{S}(r)| + A''' \sup_{r\varepsilon Z} |\hat{S}(r)| (N-M+1)^{-1}.$$

Thus if $N \geq M \geq 1$ we have

$$\left| \sum_{n=-N}^{N} \widehat{\phi S}(n) \exp int - \phi(t) \sum_{n=-N}^{N} \hat{S}(n) \exp int \right|$$

$$\leq A_0 \left(\sup_{|r|\geq M} |\hat{S}(r)| + (N-M)^{-1} \sup_{r\varepsilon Z} |\hat{S}(r)| \right) .$$

For $N \geq 100$ we can choose M so that $3N/4 \geq M \geq N$ and so

$$\left| \sum_{n=-N}^{N} \phi S(n) \exp int - \phi(t) \sum_{n=-N}^{N} \hat{S}(n) \exp int \right|$$

$$\leq A_0 \left(\sup_{|r|\geq N/2} |\hat{S}(r)| + 4N^{-1} \sup_{r\varepsilon \mathbb{Z}} |\hat{S}(r)| \right) \to 0$$

uniformly as $N \to \infty$. □

Finally we have the lemma involving the "missing link."

<u>Lemma 5.6</u>. Suppose $a_j \varepsilon \mathbb{C}$ and there exists a constant K such that $\left| \sum_{r=-N}^{N} a_r \exp irt \right| \leq K$ for all $t \varepsilon T$ and all $N \geq 1$. Then $\sum_{r=-N}^{N} a_r \exp irt$ converges almost everywhere as $N \to \infty$.

Proof. By Parsevals theorem or direct computation

$$\sum_{r=-N}^{N} |a_r|^2 = \frac{1}{2\pi}\int_{\mathbb{T}} |\sum_{r=-N}^{N} a_r \exp irt|^2 dt \leq K^2 \quad \text{for all} \quad N .$$

Thus $\sum_{r=-\infty}^{\infty} |a_r|^2$ converges and $(a_r) \in \ell^2$. Hence by Carleson's theorem (the new, powerful, ingredient) $\sum_{-N}^{N} a_r \exp irt$ converges almost everywhere as $N \to \infty$. □

Proof Theorem 5.1. Suppose, if possible, we could find (a_r) with $a_n \to \infty$ as $|n| \to \infty$ and $\sum_{r=-N}^{N} a_r \exp irt$ everywhere boundedly divergent. By Lemma 5.2 we can find an interval J on which $\sum_{r=-N}^{N} a_r \exp irt$ is uniformly bounded. By Corollary 5.4 we can find (b_r) such that $\sum_{r=-N}^{N} b_r \exp irt$ is uniformly bounded on $\mathbb{T}$ but $\sum_{-N}^{N} b_r \exp irt$ diverges at each $t \in J$ (since $\sum_{-N}^{N} a_r \exp irt$ does). This contradicts Lemma 5.6 so Theorem 5.1 follows by reduction and absurdum. □

§6. More on Sets of Convergence.

The reasoning of section 5 gives slightly more with no extra trouble.

Lemma 6.1. Suppose $a_r \to 0$ as $|r| \to \infty$. If there is an interval $I \subset \mathbb{T}$ such that $\sum_{r=-N}^{N} a_r \exp irt$ is bounded for each $t \in I$ then we can find an interval $J \subseteq I$ such that $\sum_{r=-N}^{N} a_r \exp irt$ converges almost everywhere on I as $N \to \infty$. □

Proof. As for Theorem 5.1. (Notice that the example $a_r = 1$ for $r \geq 0$, $a_r = 0$ for $r < 0$ where $\sum_{r=0}^{N}$ expirt is bounded on $(\pi/2, 3\pi/2)$ but converges nowhere shows that the condition $a_r \to 0$

as $|r| \to \infty$ cannot be dropped).

This remark is the basis for the proof of our last theorem. (The case when $a_r = \hat{f}(r)$ and $f \in L^1$ is due to me but the general result was first obtained by Lukasenko [see Zentralblatt 386-42002].)

Theorem 6.2. There exists a G_δ set such that no sequence $\sum_{-N}^{N} a_r \exp irt$ can converge on it and diverge off it (as $N \to \infty$).

Thus the set on which a power series (or more generally a trigonometric series) can converge is subject to restrictions which are not purely topological.

We need three preliminary lemmas all of which the reader can omit as trivial.

Lemma 6.3. Given an interval $[a,b]$ and $0 \leq \varepsilon < 1$ we can find a closed nowhere dense set $E \subset [a,b]$ with Lebesgue measure $\mu(E) = \varepsilon(b-a)$.

Proof. This is a trivial and well known modification of the construction of the 1/3 Cantor set. We start with $J_{01} = [a,b]$. At the nth step we have 2^{n-1} disjoint closed intervals $J_{(n-1)1}, J_{(n-1)2}, \ldots, J_{(n-1)2^n}$ of total length $(\varepsilon + 2^{-n+1}(1-\varepsilon))(b-a)$. We construct 2^n intervals J_{nr} as follows. If $J_{(n-1)m} = [x,y]$ then $J_{n(2m-1)} = [x, x+\eta_n(x-y)]$, $J_{n(2m)} = [y-\eta_n(x-y), y]$ with $\eta_n = (\varepsilon + 2^{-n}(1-\varepsilon))/2(\varepsilon + 2^{-n+1}(1-\varepsilon))$ $[1 \leq m \leq 2^{n-1}]$. It is easy to see that then J_{nr} are disjoint closed intervals of total length $(\varepsilon + 2^{-n}(1-\varepsilon))(b-a)$.

Setting $E_n = \bigcup_{r=1}^{2^n} J_{nr}$ we see that the E_n are closed and $[a,b] = E_0 \supseteq E_1 \supseteq \ldots$. Thus $E = \bigcap_{n=1}^{\infty} E_n$ is closed, $[a,b] \supseteq E$ and more generally $E_j \supseteq E$. Since E_j contains no intervals of length greater than $2^{-j}(b-a)$ we see that E contains no intervals and so is nowhere dense. Since $E_j = (\varepsilon + 2^{-j}(1-\varepsilon))(b-a)$ and $E_j \to E$ setwise, we have $\mu E = \varepsilon(b-a)$. □

<u>Lemma 6.4</u>. We can find $E_1, E_2, \ldots$ closed nowhere dense subsets such that for each interval I

$$\mu(I \cap \bigcup_{j=1}^{\infty} E_j) > 0$$

yet $\mu(\bigcup_{j=1}^{\infty} E_j) < \pi$

<u>Proof</u>. It suffices to prove the result for intervals of the form $[\alpha\pi, \beta\pi]$ with $\alpha, \beta \in \mathbb{Q}$. For then given any interval I we can find a $J = [\alpha\pi, \beta\pi]$ with $\alpha, \beta \in \mathbb{Q}$ and $I \supseteq J$ so that

$$\mu(I \cap \bigcup_{j=1}^{\infty} E_j) > 0 .$$

But the rational intervals $[\alpha\pi, \beta\pi]$ with $\alpha, \beta \in \mathbb{Q}$ can be enumerated as $J_1, J_2, \ldots$ say and by the previous lemma we can find nowhere dense closed sets $E_n \subseteq J_n$ with $\mu(E_n \cap J_n) > 0$ and $\mu(E_n) \leq 2^{-n-2}\pi$. Since $\mu(\bigcup_{j=1}^{\infty} E_j) \leq \sum_{j=1}^{\infty} \mu(E_j) < \pi$ we are done. □

Our next result is very trivial indeed.

<u>Lemma 6.5</u>. (i) If $(\lambda,\mu) \neq (0,0)$ then for any $\varepsilon > 0$

$$\mu\{t : |\lambda \exp int + \exp -int| < \varepsilon\} = \delta(\varepsilon) \qquad [n \neq 0]$$

where $\delta(\varepsilon)$ is independent of n and $\delta(\varepsilon) \to 0$ as $\varepsilon \to 0$.

(ii) If $\sum_{-N}^{N} a_n \exp int$ converges on a set of positive measure then $a_n \to 0$ as $|n| \to \infty$.

Proof. That $\mu\{t : |\lambda \exp int + \exp -int| < \varepsilon\}$ is independent of n follows from scaling. To see that $\delta(\varepsilon) \to 0$ observe then $\lambda \exp int + \exp int = A\cos nt + iB\sin nt$ with A, B real and $\max(|A|,|B|) \geq 2^{-1}\max(|\mu|,|\lambda|)$ and that

$$\mu\{t : |A\cos nt| < \varepsilon\} \to 0$$

$$\mu\{t : |B\sin nt| < \varepsilon\} \to 0 \qquad \text{as } n \to \infty.$$

(ii) If $\sum_{-N}^{N} a_n \exp int$ converges on a set E of positive measure $a_N \exp iNt + a_{-N} \exp iNt \to 0$. Thus for each $\varepsilon > 0$ $E = \bigcup_{N=1}^{\infty} E_{N\varepsilon}$ where $E_{N\varepsilon} = \{t : |a_M \exp iMt + a_{-M} \exp iMt| < \varepsilon$ for all $M \geq N\}$. Thus for some $M(0)$, $\mu(E_{N(0)12}) \geq \mu(E)/2$. Now use (i). □

Proof of Theorem 6.1. Take $E_1, E_2, \ldots$ are in Lemma 6.4. Then $\mathbf{T} \setminus \bigcup_{j=1}^{\infty} E_j$ is a G_δ set. We claim that no series $\sum_{n=-N}^{N} a_n \exp int$ can converge on $\mathbf{T} \setminus \bigcup_{j=1}^{\infty} E_j$ and diverge off it.

For suppose $\sum_{n=-N}^{N} a_n \exp int$ did diverge for all $t \in \bigcup_{j=1}^{\infty} E_j$ and converge for all $t \notin \bigcup_{j=1}^{\infty} E_j$. Since $\mu(\mathbf{T} \setminus \bigcup_{j=1}^{\infty} E_j) > 2\pi - \pi > 0$ we know that $a_n \to 0$ as $|n| \to \infty$ and so we are free to employ Lemma 6.1. Take $\mathbf{T} = I_0$. By Lemma 6.1 either $\sum_{-N}^{N} a_n \exp int$ is unbounded for some $t \in I_0$ or there exists a subinterval $J \subseteq I_0$ such that $\sum_{n=-N}^{N} a_n \exp int$ converges almost everywhere on J. But by Lemma 6.4 $\mu(J \cap \bigcup_{j=1}^{\infty} E_j) > 0$ so $\sum_{n=-N}^{N} a_r \exp int$ diverges on a subset of J of positive measure. Thus $\sum_{n=-N}^{N} a_n \exp int$ is unbounded for some $t_0 \in I_0$. It follows that we can find an $N(0)$ such that $|\sum_{n=-N(0)}^{N(0)} a_n \exp int_0| \geq 2$. By continuity we can find a subinterval $I_0' \subseteq I_0$ with $t_0 \in I_0'$ such that $|\sum_{n=-N(0)}^{N(0)} a_n \exp int| \geq 1$ for all $t \in I_0'$. Now E_1 is nowhere dense so we can find a closed subinterval $J_1 \subseteq I_0'$ with $J_1 \cap E_1 = \emptyset$.

Exactly the same arguments show that there must be a $t_1 \varepsilon I_1$ such that $\sum_{n=-N}^{N} a_n \exp int$ is unbounded. Thus we can find an $N(1)$ with $|\sum_{n=-N(1)}^{N(1)} a_n \exp int| \geq 4$, a subinterval $I_1' \subseteq I_1$ with $|\sum_{n=-N(1)}^{N(1)} a_n \exp int| \geq 2$ for all $t \varepsilon I_1'$ and a closed subinterval $J_2 \subseteq I_1'$ with $J_2 \cap E_2 \neq \emptyset$. Continuing we find a sequence of closed intervals $J_1 \supseteq J_2 \supseteq \ldots$ and integers $N(0) \leq N(1) \leq \ldots.$ such that

$(i)_k \quad J_k \cap E_k = \emptyset$

$(ii)_k \quad |\sum_{n=-N(k-1)}^{N(k-1)} a_n \exp int| \geq 2^{k-1} \quad$ for all $t \varepsilon J_k$.

Since the J_k are nested closed sets $\bigcap_{k=1}^{\infty} J_k \neq \emptyset$. Choose $x \varepsilon \bigcap_{k=1}^{\infty} J_k$. By $(i)_k$) $x \notin \bigcup_{j=1}^{\infty} E_j$ and by $(ii)_k$

$$\sup_N |\sum_{r=-N}^{N} a_n \exp inx| = \sup_k |\sum_{n=-N(k)}^{N(k)} a_n \exp inx| = \infty.$$

Thus $\sum_{n=-N}^{N} a_n \exp int$ diverges for some $t \notin \bigcup_{j=1}^{\infty} E_j$ and the result follows by contradiction. □

There are at least 3 natural questions one can ask at this point but I know the answer to none of them.

(1) The counter examples we have just constructed involve sets of positive measures. Is the following statement true or false? "If X is a $G_{\delta\sigma}$ set of Lebesgue measure zero can we find a_r with $\sum_{n=-N}^{N} a_n \exp int$ divergent on X and convergent on $\mathbb{T} \setminus X$".

(2) Can we obtain Theorem 6.2, at least in the case of power series (i.e. $a_r = 0$ for $r < 0$), by complex variable means?

(3) In view of Theorem 5.1, what, if anything, can we say about $\{t : \sum_{n=-N}^{N} a_n \exp int$ diverges boundedly as $N \to \infty\}$ (However I suspect that the answer to (3) may be a mess involving Hausdorff measures as well as topology and Lebesgue measure.)

$L^p - L^q$ MAPPING PROPERTIES OF THE RADON TRANSFORM

Daniel M. Oberlin*

For $1 \leq p \leq \infty$ write $L^p(R^2)$ for the usual Lebesgue space formed from two-dimensional Lebesgue measure on R^2 and write $L^p(D)$ for the subspace of functions in $L^p(R^2)$ supported in the closed unit disc D. Let $L^p(T\times R^+)$ be the Lebesgue space formed from the product of one-dimensional Lebesgue measures on $T = [0,2\pi)$ and $R^+ = [0,\infty)$. The Radon transform Rf of $f \in L^1(R^2)$ is the function $Rf(\theta,r)$ on $T\times R^+$ defined by

$$Rf(\theta,r) = \int_{-\infty}^{\infty} f(r\cos\theta - t\sin\theta,\ r\sin\theta + t\cos\theta)\ dt\,.$$

The references given in the first two sections of [2] might form a convenient point of departure for the reader caring to pursue the extensive literature on the Radon transform and its applications. Here we are interested in the following question.

For what values of p and q does the Radon transform map $L^p(R^2)$ or $L^p(D)$ into $L^q(T\times R^+)$?

It is easy to see that $RL^1(R^2) \subseteq L^1(T\times R^+)$ and so $RL^1(D) \subseteq L^1(T\times R^+)$. Since $f \in L^p(D)$ implies that Rf is supported in $T\times[0,1]$, and since L^q spaces formed from a finite measure decrease as q increases, we can interpret the inclusion $RL^p(D) \subseteq L^q(T\times R^+)$ for $1 < p < q < \infty$ as evidence that the averaging operator R improves the behavior of functions in $L^p(D)$. The extent to which this occurs is part of the content of the following theorem.

<u>Theorem</u>. With reference to figure 1 below,

(a) if $RL^p(D) \subseteq L^q(T\times R^+)$, then (p^{-1},q^{-1}) lies in the closed quadrilateral OABC;

---

*Partially supported by NSF Grant MCS-7827602.

(b) if (p^{-1},q^{-1}) is a point different from A in the closed quadrilateral OABC, then $RL^p(D) \subseteq L^q(T\times R^+)$;

(c) if $RL^p(R^2) \subseteq L^q(T\times R^+)$, then (p^{-1},q^{-1}) lies on the closed segment AB;

(d) if (p^{-1},q^{-1}) is a point different from A on the closed segment AB, then there is a constant $K(p,q)$ such that

$$\left(\int_{T\times R^+} |Rf|^q\right)^{1/q} \leq K(p,q) \left(\int_{R^2} |f|^p\right)^{1/p}$$

holds for $f \in L^1 \cap L^p(R^2)$.

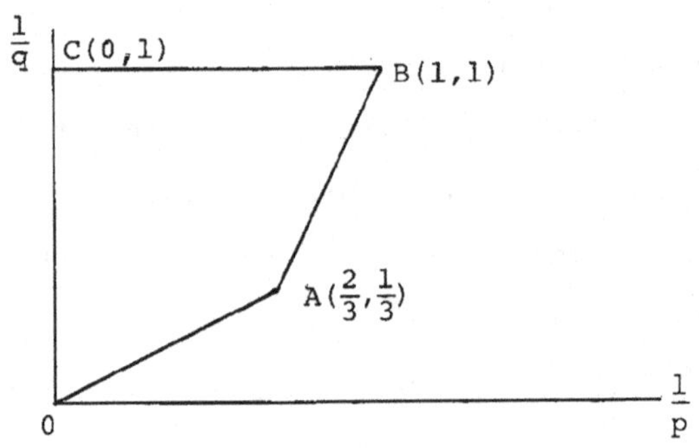

Figure 1

Remark: It can be shown (by a method more sophisticated but less transparent than the one used here) that R maps $L^{3/2}(R^2)$ into $L^3(T\times R^+)$. For this and related results, see [0].

Proof: We begin by establishing (a) and (c). For a function f on R^2, $||f||_p$ will denote the norm of f in $L^p(R^2)$ and $||Rf||_q$ will stand for the norm of Rf in $L^q(T\times R^+)$. If f_r is the characteristic function of a disc of radius r centered at the origin, then

$$\|f_r\|_p = \pi^{1/p} r^{2/p}$$

while $\|Rf_r\|_q$ is on the order of $r^{1+1/q}$. Thus if

$$\text{(1)} \qquad RL^p(D) \subseteq L^q(T\times R^+)$$

we must have

$$\text{(2)} \qquad 1 + \frac{1}{q} \geq \frac{2}{p}.$$

(Else for suitably chosen sequences $r_n \to 0$ and $a_n > 0$ we would have

$$f = \sum a_n f_{r_n} \in L^p(R^2)\ ,\quad Rf \notin L^q(T\times R^+)\ .)$$

Similarly, if g_ε is the characteristic function of a rectangle centered at the origin with sides 1/2 and ε, then

$$\|g_\varepsilon\|_p = (\varepsilon/2)^{1/p}$$

while $\|R_{g\varepsilon}\|_q$ is on the order of $\varepsilon^{2/q}$. As above, (1) implies that $2/q \geq 1/p$. With (2) this establishes (a). Now consider $L^p(R^2)$ instead of $L^p(D)$. If $1+1/q > 2/p$, then we can construct functions $f = \sum a_n f_{r_n}$ with $r_n \to \infty$ such that $f \in L^p(R^2)$, $Rf \notin L^q(T\times R^+)$. With (a) this yields (c).

The proofs of (b) and (d) will depend on the extended Marcinkiewicz interpolation theorem (Theorem 3.15, p. 197, [1]). Since R is a bounded mapping of $L^1(R^2)$ into $L^1(T\times R^+)$ and of $L^\infty(D)$ into both $L^\infty(T\times R^+)$ and $L^1(T\times R^+)$, it is only necessary to show that R is of restricted weak type (3/2,3) when considered as a mapping of functions on R^2 to functions on $T\times R^+$. Thus, letting λ be Lebesgue measure on R^2 and μ be the product measure on $T\times R^+$, we must show that the inequality

$$s^{\frac{3}{2}}\,[\mu\,(R\chi_E > s)]^{\frac{1}{2}} \le c\lambda(E) \tag{3}$$

holds for some constant c, all $s > 0$, and all measurable $E \subseteq R^2$. (Here χ_E is the characteristic function of E.) We begin with some reductions.

First, the trivial inequality

$$\|Rf\|_1 \le \pi\|f\|_1 \tag{4}$$

implies that $s\mu(R\chi_E > s) \le \pi\lambda(E)$. If also $s < 3\mu(R\chi_E > s)/2\pi$, then (3) holds with $c = \left(\frac{3\pi}{2}\right)^{1/2}$. Thus we can assume that

$$s \ge \frac{3}{2\pi}\,\mu(R\chi_E > s). \tag{5}$$

Next, it follows from (4) that if $E_1 \supseteq E_2 \supseteq \cdots$ is a decreasing sequence of sets of finite measure and if (3) holds with E replaced by E_n for each n, then (3) holds for $\bigcap E_n$. We can therefore assume that E is open.

So let $E \subseteq R^2$ be an open set of finite measure, and we will establish (3). Let $\{f_n\}$ be a sequence of continuous compactly supported functions on R^2 such that

$$0 \le f_n \le \chi_E,\quad f_n(x) \uparrow \chi_E(x) \quad \text{for all } x \in R^2.$$

Then

$$\{Rf_n > s\} \subseteq \{R\chi_E > s\},$$

and the union of the sets on the left is equal to the set on the right. Since each Rf_n is continuous on R^2 (we identify R^2 and $T\times R^+$ in the usual way by identifying $(r\cos\theta, r\sin\theta) \in R^2$ with $(\theta, r) \in T\times R^+$), there is an open subset G of $\{R\chi_E > s\}$, which we can take to the union of a finite number of open discs, such that

$\mu(G) \geq \mu(R\chi_E > s)/2$. Putting

$$\gamma(\theta) = \int_0^\infty \chi_G(r\cos\theta,\ r\sin\theta)\,dr,$$

we see that $\gamma(\theta)$ is a continuous function of $\theta \in R$ and

(6) $$\mu(R\chi_E > s)/2 \leq \int_0^{2\pi} \gamma(\theta)\,d\theta = \mu(G) \leq \mu(R\chi_E > s)\ .$$

Now for $\theta \in R$ define

$$E(\theta) = \{x = (r\cos\theta - t\sin\theta, r\sin\theta + t\cos\theta) \in R^2 :$$

$$r \in [0,\infty), (r\cos\theta, r\sin\theta) \in G,\ t \in R,\ x \in E\},$$

so that, in particular, $E(\theta) \subseteq E$.

We need two facts whose proofs will be postponed:

(7) $$\lambda(E(\theta)) \geq s\gamma(\theta)\ ,$$

(8) if $|\theta_1 - \theta_2| \leq \frac{\pi}{2}$, then $\lambda(E(\theta_1) \cap E(\theta_2)) \leq \dfrac{\pi\gamma(\theta_1)\gamma(\theta_2)}{2|\theta_1 - \theta_2|}$.

With (7) and (8) and if $|\theta_j - \theta_k| \leq \pi/2$ for $j,k = 1,2,\ldots,n$, we can estimate $\lambda(E)$ as follows:

(9) $$\lambda(E) \geq \sum_1^n \gamma(E(\theta_j)) - \sum_{1\leq j<k\leq n} \lambda(E(\theta_j) \cap E(\theta_k)) \geq$$
$$s \sum_1^n \gamma(\theta_j) - \frac{\pi}{2} \sum_{1\leq j<k\leq n} \frac{\gamma(\theta_j)\gamma(\theta_k)}{|\theta_j - \theta_k|}\ .$$

Next we choose n and $\theta_1, \ldots, \theta_n$. By (5) and (6) we have $s \geq 3 \int_0^{2\pi} \gamma(\theta)\,d\theta/2\pi$, so there is an $n = 1, 2, \ldots$ such that

(10) $$s_1 \doteq \frac{3n^2}{2\pi} \int_0^{2\pi} \gamma(\theta)\,d\theta \leq s < \frac{3(n+1)^2}{2\pi} \int_0^{2\pi} \gamma(\theta)\,d\theta\ .$$

Let

$$0 = \theta_1' < \theta_2' < \ldots < \theta_n' = \frac{\pi}{2}$$

be equispaced points, and choose $\theta_0 \varepsilon R$ such that

$$\sum_1^n \gamma(\theta_0 + \theta_j') = \frac{n}{2\pi} \int_0^{2\pi} \gamma(\theta)\, d\theta \, .$$

This is possible since γ is continuous. Taking θ_j to be $\theta_0 + \theta_j'$ $(1 \leq j \leq n)$ and taking account of the fact that

$|\theta_j - \theta_k| \geq \pi/2n$ if $1 \leq j < k \leq n$, (9) yields

$$\lambda(E) \geq \frac{sn}{2\pi} \int_0^{2\pi} \gamma(\theta)\, d\theta - n \sum_{1 \leq j < k \leq n} \gamma(\theta_j)\gamma(\theta_k) \geq$$

$$\frac{sn}{2\pi} \int_0^{2\pi} \gamma(\theta)\, d\theta - n^3 \left[\frac{1}{n} \sum_1^n \gamma(\theta_j)\right]^2 \geq$$

$$\frac{s_1 n}{2\pi} \int_0^{2\pi} \gamma(\theta)\, d\theta - n^3 \left[\frac{1}{2\pi} \int_0^{2\pi} \gamma(\theta)\, d\theta\right]^2 .$$

Since, by (10),

$$n^2 = 2\pi s_1/3 \int_0^{2\pi} \gamma(\theta)\, d\theta \, ,$$

we have

$$\lambda(E) \geq (2\pi)^{-1/2}(3^{-1/2} - 3^{-3/2})\, s_1^{3/2} \left[\int_0^{2\pi} \gamma(\theta)\, d\theta\right]^{1/2} .$$

But, by (10) and (6),

$$s_1 \geq \left(\frac{n}{n+1}\right)^2 s \, , \int_0^{2\pi} \gamma(\theta)\, d\theta \geq \mu(R\chi_E > s)/2 \, .$$

Thus

$$\lambda(E) \geq \pi^{-1/2} 2^{-4} \, (3^{-1/2} - 3^{-3/2})\, s^{3/2} \, [\mu(R\chi_E > s)]^{1/2} \, ,$$

and this gives (3). It remains to establish (7) and (8).

<u>Proof of (7)</u>: By the definition of $E(\theta)$,

$$\lambda(E(\theta)) = \int_0^\infty \chi_G(r\cos\theta, r\sin\theta) \int_{-\infty}^{\infty} \chi_E(r\cos\theta - t\sin\theta,$$

$$r\sin\theta + t\cos\theta)\,dt\,dr =$$

$$\int_0^\infty \chi_G(r\cos\theta, r\sin\theta)\, R\chi_E(\theta, r)\,dr .$$

Since $G = \{R\chi_E > s\}$ and since $\gamma(\theta) = \int_0^\infty \chi_G(r\cos\theta, r\sin\theta)\,dr$, (7) is evident.

<u>Proof of (8)</u>: For a subset S of $[0,\infty)$, let

$$F(\theta, S) = \{(r\cos\theta - t\sin\theta, r\sin\theta + t\cos\theta) \in R^2 :$$

$$r \in S, t \in R\} .$$

If $\theta \in R$, put

$$R(\theta) = \{r \in [0,\infty) : (r\cos\theta, r\sin\theta) \in G\} .$$

Then

$$E(\theta) = F(\theta, R(\theta)) .$$

Now, for $j = 1,2, R(\theta_j)$ is an open subset of $[0,\infty)$ which we can write as the union of disjoint intervals I_ℓ^j: $R(\theta_j) = \bigcup_{\ell=1}^{\infty} I_\ell^j$. Thus

$$\lambda(E(\theta_1) \cap E(\theta_2)) \leq \sum_{k,\ell=1}^{\infty} \lambda(F(\theta_1, I_k^1) \cap F(\theta_2, I_\ell^2)) .$$

If ε_ℓ^j denotes the length of I_ℓ^j and if we recall that

$$\gamma(\theta_j) = \int_0^\infty \chi_G(r\cos\theta_j, r\sin\theta_j)\,dr = \sum_{\ell=1}^{\infty} \varepsilon_\ell^j ,$$

then (8) follows when we observe that the parallelogram $F(\theta_1, I_k^1) \cap F(\theta_2, I_\ell^2)$ has area $\varepsilon_k^1 \varepsilon_\ell^2 / |\sin(\theta_1 - \theta_2)|$.

References

[0] D. Oberlin and E. Stein, Mapping properties of the Radon trans form. Indiana Univ. Math. J. 31 [1982], to appear.

[1] E. Stein and G. Weiss, Introduction to Fourier analysis on Euclidean spaces. Princeton University Press, Princeton, NJ, 1971.

[2] L. Zalcman, Offbeat integral geometry, Amer. Math. Monthly 87 (1980), 161-175.

QUALITATIVE RATIONAL APPROXIMATION ON PLANE COMPACTA

A. G. O'Farrell*

Abstract

Let X be a compact subset of the complex plane. Let $R(X)$ denote the space of all rational functions with poles off X. Let $\mathcal{A}(X)$ denote the space of all complex-valued functions on X that are analytic on the interior of X. Let $A(X)$ be a Banach space of functions on X, with $R(X) \subset A(X) \subset \mathcal{A}(X)$. Consider the problems: (1) Describe the closure of $R(X)$ in $A(X)$. (2) For which X is $R(X)$ dense in $A(X)$? There are many results on these problems, for various particular Banach spaces $A(X)$. We offer a point of view from which these results may be viewed systematically.

1. Introduction.

(1.1) Let me begin by indicating how my topic fits into the world of mathematics.

Qualitative rational approximation theory answers the question: is it possible to approximate a given function arbitrarily closely by rationals? There is also a quantitative theory, which addresses the question: how closely can a given function be approximated by rational functions of a specified degree? Experience indicates that the quantitative theory lags behind the qualitative by anything up to thirty years. For the kind of qualitative results I shall present, the quantitative theory is nonexistent or primitive.

After the quantitative theory comes the computational theory, culminating in practical computer programs. The expected time lag here is another twenty years or thereabouts so perhaps by 2020 A.D. the results will be ready for use by the engineers and scientists. What

*Department of Mathematics, University of Connecticut, Storrs, CT 06268. Present address: Department of Mathematics, St. Patrick's College, Maynooth, County Kildare, Ireland.

will the engineers and scientists do with them? It is probably better not to know. For instance, the theorem of Weierstrass, that periodic functions on the line may be approximated uniformly by trigonometric polynomials, finds application in digital recording. This makes possible the indefinite preservation, with effectively complete fidelity, of all kinds of unspeakable rubbish.

I shall not deal explicitly with polynomial approximation. Except in the case of L^p approximation $(1 \leq p < +\infty)$, polynomial approximation is possible if and only if rational approximation is possible and the set is polynomially-convex. For L^p polynomial approximation there are very interesting and formidable problems. See [2] for results and references to the work of M. M. Džrbašjan, V. P. Havin, V. G. Mazja, S. N. Mergelyan, A. P. Tamadjan, A. L. Šaginjan, and S. O. Sinanjan.

I restrict myself to compact sets. For weighted uniform approximation the theory on closed unbounded sets has been developed by N. U. Arakelian, P. M. Gauthier, W. Hengartner, A. Roth, S. Scheinberg, J. L. Walsh, and others. See [7,8] for references.

The theory in several variables is comparatively primitive.

(1.2) My objective is to give the essential facts about qualitative rational approximation on plane compacta. I shall endeavour to make the results seem intelligible and natural, but I will not include any proofs. Suffice it to say that the proofs are quite inhomogeneous, and in many cases long and intricate.

My account will not be historical, nor will it reflect the logical structure of the existing theory. Instead, I shall present the results in an order in which I would <u>like</u> to be able to prove them, from my point of view.

The results I shall present were developed in the period 1955-1980. The main contributers to the core of the theory were T. Bagby, A. Browder, L. Carleson, A. M. Davie, E. P. Dolženko, V. P. Havin,

S. Ya. Havinson, J. Garnett, A. A. Gonchar, L. I. Hedberg, M. S. Melnikov, S. N. Mergelyan, S. O. Sinanjan, A. G. Vitushkin, and J. Wermer. In the prehistory of the subject, during the first seventy years after the Weierstrass approximation theorem, the outstanding contributers were C. Runge, H. Lebesgue, J. L. Walsh, T. Carleman, A. Roth, O. J. Farrell, F. Hartogs, A. Rosenthal, and M. V. Keldysh.

(1.3) Arbitrary plane compacta can be pretty complicated objects, and as a result, rational approximation on such sets is an extremely perverse subject, teeming with surprising counterexamples. I am not going to dwell on examples. I am going to present positive results. In doing this, I run the risk that the beginner may not appreciate that the most surprising thing about the subject is the existence of a moderately extensive body of positive results; and consequently that he may feel that the formulation of some of these results is a little complicated. He can rest assured that all the simple answers have been tried and found wanting.

§2. Formulation of the problems.

(2.1) Let X be an arbitrary compact subset of the complex plane, $\mathbb{C}$. Let $\mathcal{R}(X)$ denote the space of all rational functions with poles off X. Let $\mathcal{A}(X)$ denote the space of all functions on X, analytic on the interior of X. Let $B(X)$ be one of a certain list (see (2.2) below) of Banach spaces of complex-valued functions on X, such that $\mathcal{R}(X) \subseteq B(X)$ and the subspace $A(X) = B(X) \cap \mathcal{A}(X)$ is closed. Let $R(X)$ denote the closure of $\mathcal{R}(X)$ in $B(X)$. Then $R(X) \subset A(X)$. We consider the following two main problems:

(1) Give a reasonable description of $R(X)$, i.e. give an explicit condition on a function $f \in B(X)$, necessary and sufficient for $f \in R(X)$.

(2) For which compact X is $R(X) = A(X)$?

(2.2) The main examples of Banach function spaces $B(X)$ are as follows.

(1) $L^p(X) = L^p(X,m)$ $(1 \leq p < \infty)$, where m is area measure on X.

(2) $C(X)$, the space of continuous functions on X, with the sup norm, $\|f\|_\infty$.

(3) $\mathrm{Lip}(\alpha,X)$ $(0 < \alpha \leq 1)$, the space of functions f for which there exists a constant $\kappa > 0$ such that

$$|f(x) - f(y)| \leq \kappa|x-y|^\alpha$$

for all $x,y \in X$. The norm of f is $\|f\|_\infty$ plus the least value of κ.

(4) $C^k(X)$, for k a positive integer. This space has two versions. Let C^k denote the space of k-times continuously-differentiable functions on $\mathbb{C}$, with bounded partials up to order k. Then C^k forms a Banach algebra with the norm

$$\|f\| = \sum_{j=0}^{k} \sum_{r+s=j} \frac{r!s!}{j!} \left\| \frac{\partial^j f}{\partial x^r \partial y^s} \right\|_\infty .$$

Let I and J denote the closed ideals

$I = \{f \in C^k : f = 0 \text{ on } X\}$,

$J = \{f \in C^k : \dfrac{\partial^j f}{\partial x^r \partial y^s} = 0 \text{ on } X,\ 0 \leq j \leq k,\ r+s=j\}$.

The <u>function version</u> of $C^k(X)$ is the quotient space C^k/I (with the quotient norm). The <u>jet version</u> of $C^k(X)$ is the quotient space C^k/J. The jet version has an alternative, local, description, via Whitney's extension theorem [23, Ch. 6]. The natural quotient map from the jet version onto the function version is occasionally injective and a homeomorphism, but usually not. In case $k = 1$, the function version has a local description [13].

(5) Lip$(k+\alpha, X)$, $0 < k \in \mathbb{Z}$, $0 < \alpha \leq 1$. This space has a function version and a jet version, derived from the global space Lip$(k+\alpha)$ of functions in C^k with k-th partial derivatives in Lip α. There is a local description of the space [23, Ch. 6, p. 176].

(6) The Sobolev spaces $W^{k,p}(X)$, $0 < k \in \mathbb{Z}$, with function and jet versions, derived from the global spaces $W^{k,p}$ of functions whose k-th distribution derivatives are representable by L^p functions.

Apart from these main examples, there are weighted L^p spaces, mean Lipschitz spaces, weighted Sobolev spaces, and so on.

Note that in every instance B is really a function $X \mapsto B(X)$ on the compact subsets of $\mathbb{C}$. Moreover, in each instance $B(X)$ may be derived from a global space $B = B(\mathbb{C})$ by restriction, and the norm induced by $B(\mathbb{C})$ is equivalent to the given norm. Thus $B(F)$ may be defined by restriction for all closed $F \subset \mathbb{C}$. Each B has a <u>localness</u> property: if X is compact, $f : X \to \mathbb{C}$, and each point $a \in X$ has a closed relative neighbourhood Y in X such that $f \in B(Y)$, then $f \in B(X)$. Each B has a technical property, called T-<u>invariance</u>. It states that if $f \in B(X)$ and $\phi \in \mathcal{D}$(= the space of C^∞ functions with compact support), then the function $T_\phi f$ defined by

$$T_\phi f = \phi f - \widehat{f \frac{\partial \phi}{\partial \bar{z}} m}$$

also belongs to $B(X)$, where the <u>Cauchy transform</u> $\hat{\mu}$ of a measure μ is defined by

$$\hat{\mu}(z) = \frac{1}{\pi} \int \frac{d\mu(\zeta)}{z-\zeta} .$$

The operator T_ϕ is called the <u>Vitushkin localization operator</u>. It is used to chop up the singular set of an analytic function.

(2.3) The above examples of Banach function spaces $B(X)$ are ordered by continuous (and usually compact) inclusion maps:

$$W^{k+1,p}(X) \to C^k(X) \to W^{k,p}(X) ,$$

$$\mathrm{Lip}(k+\alpha,X) \to C^k(X) \to \mathrm{Lip}(k-1+\alpha,X) ,$$

$$\mathrm{Lip}(\alpha,X) \to C(X) \to L^p(X) ,$$

$$L^p(X) \to L^{p'}(X) , \quad (p > p') ,$$

$$W^{k,p}(X) \to W^{k,p'}(X) , \quad (p > p') ,$$

$$W^{k+1,p}(X) \to \mathrm{Lip}(k+\tfrac{1}{q},X) , \quad (\tfrac{1}{p}+\tfrac{1}{q} = 1) ,$$

$$\mathrm{Lip}(k+\alpha,X) \to \mathrm{Lip}(k+\alpha',X) , \quad (\alpha > \alpha) .$$

Approximation in one space implies approximation in all larger spaces, since the injections are continuous.

§3. Preliminary.

(3.1) Rational approximation in all these Banach spaces is local. Precisely speaking, we have the following result, due to E. Bishop for $B = C$.

Theorem. Let $f \in B(X)$, and suppose every point $a \in X$ has a closed relative neighbourhood $Y \subset X$ such that $f|Y \in R(Y)$. Then $f \in R(X)$.

This means that the answers to the main questions (1) and (2) of (2.1) should involve only local properties of f and X. The proof is an application of the localization operator, as in [14].

(3.2) *Theorem* (Runge). *Let* $f \in B(X)$ *be (the restriction to* X *of a function) analytic on a neighbourhood of* X. *Then* $f \in R(X)$.

See [15]. As a result of this, the closure of $\mathcal{R}(X)$ in $B(X)$ is the same as the closure of

$$\mathcal{H}(X) = \{f : f \text{ is analytic on a neighbourhood of } X\}.$$

(3.3) From the C^1 level up, i.e. as soon as $B \hookrightarrow C^1$ is continuous, the point Cauchy-Riemann operator (at a point $a \in X$):

$$\bar{\partial} \cdot (a) : f \to \bar{\partial} f(a) = \tfrac{1}{2}\left\{\frac{\partial f}{\partial x}(a) + i\,\frac{\partial f}{\partial y}(a)\right\}$$

can be expected to play a rôle. It will be a continuous linear functional on $B(X)$. Since $\bar{\partial} f(a) = 0$ for $f \in \mathcal{R}(X)$, it follows by continuity that $\bar{\partial} f(a) = 0$ for $f \in R(X)$.

(3.4) Since $\mathcal{R}(X)$ consists of functions which are infinitely-differentiable on a neighbourhood of X, the closure of $\mathcal{R}(X)$ in $B(X)$ is contained in the closure of C^∞ in $B(X)$. Thus it makes little sense to work with a space $B(X)$, in which C^∞ is not dense. This is why $L^\infty(X)$ was left out of the list in (2.2); the closure of C^∞ in $L^\infty(X)$ is $C(X)$. For the same reason we must work with the spaces $\mathrm{lip}(k+\alpha, X)$ instead of $\mathrm{Lip}(k+\alpha, X)$, for $0 < \alpha < 1$. These spaces are the closures of C^∞ in the respective norms. For $0 < \alpha < 1$, $\mathrm{lip}(\alpha, X)$ is the space of functions $f : X \to \mathbb{C}$ such that, given $\varepsilon > 0$, there exists $\delta > 0$ such that

$$|f(x) - f(y)| \leq \varepsilon |x - y|^\alpha$$

whenever $x \in X$, $y \in X$, and $|x - y| < \delta$. It is also the space of restrictions of the corresponding global $\mathrm{lip}\,\alpha$ space [23]. For $0 < k \in \mathbb{Z}$ and $0 < \alpha < 1$, $\mathrm{lip}(k+\alpha, X)$ is the space (with two versions) of functions (or jets) on X with extensions in global $\mathrm{lip}(k+\alpha)$, which in turn is the space of functions $f : \mathbb{C} \to \mathbb{C}$ such that f and all partial derivatives up to order k belong to $\mathrm{lip}\,\alpha$. For $0 < k \in \mathbb{Z}$, the closure of C^∞ in $\mathrm{Lip}(k,X)$ is not so easily described, but for $k = 1$, it is known [16] that

$$f \in \mathrm{clos}_{\mathrm{Lip}(1,X)} R(X)$$

if and only if

$$f \in \mathrm{clos}_{C^1(X)} R(X) ,$$

i.e. Lip 1 rational approximation reduces to C^1 rational approximation; this renders it unnecessary to describe the Lip 1 closure of C^∞ on X. It may be possible to do the same thing for Lip k.

(3.5) In general, an optimist might hope that the answer to main problem (2) would be categorically appropriate to the functor B, and would be free of "analytic" elements, i.e. that for $B = L^p$ it would involve only area, for $B = C$ it would be topological, for $B = \mathrm{Lip}\,\alpha$ it would be metric, for $B = C^1$ it would involve the C^1 - differential structure of X, and so on. As we shall see, some of these hopes are fulfilled.

(3.6) The outline of the theory is as follows. The L^p theory, for $1 \leq p < 2$, is trivial. The remaining B divide into two broad classes: the <u>smooth</u> class $(B \hookrightarrow C^1)$, and the <u>hairy</u> class $(C^1 \hookrightarrow B\ ,\ B \neq C^1)$. In the smooth class the point Cauchy-Riemann operator plays a crucial role, and the results are pretty simple to state. In the hairy class, which embraces $L^p (2 \leq p < \infty)$, C, and $\mathrm{lip}\,\alpha (0 < \alpha < 1)$, it takes a bit of effort to digest the statements, let alone the proofs.

(3.7) There are several unsolved problems in the theory, and I have indicated some of them as they arise.

4. The Smooth Class.

(4.1) Consider the jet spaces $B = C^k$, lip(k+α). All jets in R(X) satisfy the Cauchy-Riemann equations on X. The jets in A(X) need only satisfy the Cauchy-Riemann equations on the closure of the interior of X. For any point $a \in X \sim \text{clos int } X$, there exists $f \in A(X)$ such that $\bar{\partial} f(a) \neq 0$. Thus, if $X \neq \text{clos int } X$, then $R(X) \neq A(X)$.

The following theorem solves main problem (1) for the jet space $B = \text{lip}(k+\alpha)$ [17].

Theorem. Let $0 < k \in \mathbb{Z}$, $0 < \alpha < 1$, $B = \text{lip}(k+\alpha)$, and let $X \subset \mathbb{C}$ be compact. Then R(X) is the set of all jets $f \in B(X)$ such that

$$\frac{\partial^j \bar{\partial} f}{\partial x^r \partial y^s} = 0$$

on X for $0 \le j \le k-1$, $r + s = j$.

This says that a jet is approximable by rationals if and only if it satisfies the Cauchy-Riemann equations on X, and also satisfies all consequences of the Cauchy-Riemann equations which make sense for jets in B(X).

As an immediate corollary, we have the solution of main problem (2) for $B = \text{lip}(k+\alpha)$.

Corollary. $R(X) = A(X)$ if and only if $X = \text{clos int } X$.

The corresponding results for C^k and $W^{k,p}$ have not yet been established.

(4.2) For the function spaces C^k and lip(k+α), problem (1) has not been solved. Problem (2) has been solved only for sets with empty interior.

<u>Theorem</u> [18,17]. <u>Let</u> $B = C^k$ or $\mathrm{lip}(k+\alpha)$, $0 < k \in \mathbb{Z}$, $0 < \alpha < 1$. <u>Then</u> $R(X) = B(X)$ <u>if and only if</u> X <u>is a subset of a finite union of pairwise-disjoint simple curves of class</u> B (i.e. class C^k if $B = C^k$ and class $\mathrm{lip}(k+\alpha)$ if $B = \mathrm{lip}(k+\alpha)$).

The corresponding result for $W^{k,p}$ has not been proved.

(4.3) I remark parenthetically that Theorem (4.1) yields the solution of the rational approximation problems in the Fréchet jet space $C^\infty(X)$, since $C^\infty(X)$ is the injective limit of the spaces $\mathrm{lip}(k+\frac{1}{2},X)$.

§5. An example.

Before presenting the results on the hairy class of B's, I give an example to indicate the flavour.

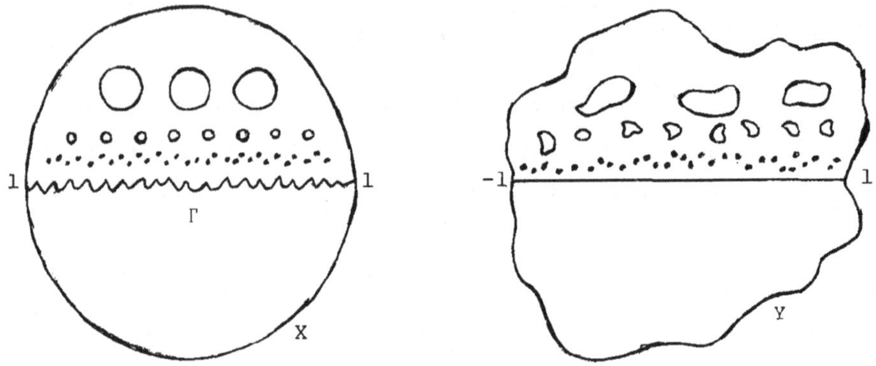

<u>Figure 1</u>

Consider the open unit disc D, and let Γ be an arc of positive area joining -1 to 1 and otherwise lying in D. Take a countable family of open discs $D_n \subset D \sim \Gamma$, with $\sum_n \operatorname{diam} D_n < +\infty$, such that each point of Γ belongs to $\operatorname{clos} \bigcup_n D_n$. Let

$$X = \text{clos } D \sim \bigcup_{n=1}^{\infty} D_n .$$

Find a homeomorphism ϕ of $\mathbb{C}$ onto $\mathbb{C}$ which maps Γ onto the line segment $[-1,1]$. Let $Y = \phi(X)$. Then it can be shown [cf. 5, p. 220 and p. 235 (13.2)] that for $B = C$ we have $R(X) \neq A(X)$ and $R(Y) = A(Y)$. Thus uniform rational approximation is not determined by topological criteria.

§6. Second main problem for the hairy class.

(6.1) For $1 \leq p < 2$, the L^p problems have a trivial answer.

Theorem. *Let* $1 \leq p < 2$, $B = L^p$. *Then* $A(X) = R(X)$ *for all compact* X.

(6.2) Consider $B = L^p$ $(2 \leq p < \infty)$, C, or $\text{lip}\,\alpha(0 < \alpha < 1)$. Problem (2) is about the relative size of the spaces $\mathcal{R}(X)$ and $A(X)$. In view of Runge's theorem, we can replace $\mathcal{R}(X)$ by $\mathcal{H}(X)$, and so the problem concerns the approximation of functions with singularities on $\mathbb{C} \sim \text{int } X$ by functions with singularities on $\mathbb{C} \sim X$. Recalling that the approximation problem is local, we fix an open disc D, and look for ways of comparing the set of functions with singularities in $D \sim \text{int } X$ with the set of functions with singularities in $D \sim X$.

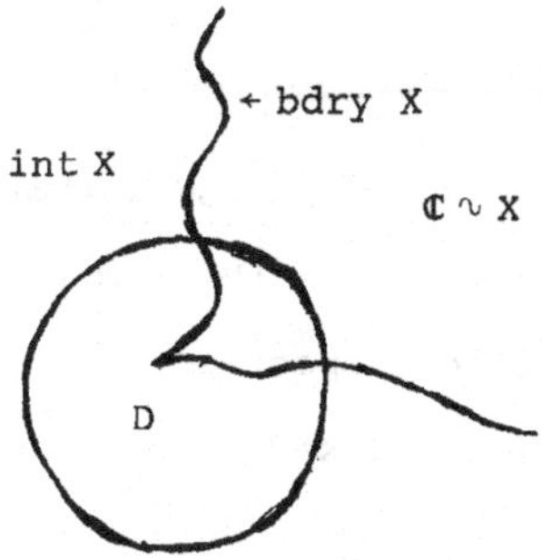

Figure 2

The key idea is to measure the relative sizes of these singularity sets by means of a capacity. A capacity is a nonnegative increasing function

$$\gamma : 2^{\mathbb{C}} \to [0,\infty].$$

The capacity γ_B associated with B assigns a number $\gamma_B(E)$ to each set $E \subset \mathbb{C}$, and this number is a measure of the size of the collection of functions with singular support in E. The definition of $\gamma_B(E)$ for compact E is

$$\sup_f |a_1(f)|,$$

where f runs over all functions in the unit ball of $A(\mathbb{C} \sim E)$ (interpreted in the obvious way, since $\mathbb{C} \sim E$ is not compact), and $a_1(f)$ is the coefficient of $1/z$ in the expansion

$$f(z) = a_0 + \frac{a_1}{z} + \frac{a_2}{z^2} + \ldots$$

of f near ∞. For arbitrary sets E, $\gamma_B(E)$ is defined as

$$\sup\{\gamma_B(F) : F \subset E, \ F \text{ compact}\}.$$

It is not very surprising that if $R(X) = A(X)$, then $\gamma_B(D \sim X) = \gamma_B(D \sim \text{int}\, X)$ for every open disc D. The extraordinary thing is that the converse holds, so that a simple collection of numerical invariants characterizes rational approximation.

<u>Theorem</u> [24,9,1,10,11,19]. <u>Let</u> $B = L_p (2 \leq p < \infty)$, C, <u>or</u> lip $\alpha (0 < \alpha < 1)$. <u>Then</u> $R(X) = A(X)$ <u>if and only if</u> $\gamma_B(D \sim \text{int}\, X) = \gamma_B(D \sim X)$ <u>for all open discs</u> D.

(6.3) For practical purposes, the above theorem is of no use without a computational description of γ_B. In order to use the theorem to decide whether or not $R(X) = A(X)$ for a specific compact set X, we need to be able to compute $\gamma_B(E)$. The problem is markedly simplified by the fact that the capacity condition of the theorem is equivalent to the formally weaker condition:

<u>There exists</u> $\kappa > 0$ <u>such that</u> $\gamma_B(D \sim X) \geq \kappa\, \gamma_B(D \sim \text{int } X)$ <u>for each open disc</u> D.

This means that it is sufficient to identify γ_B up to multiplicative bounds, i.e. to find an explicitly-computable capacity $F : 2^{\mathbb{C}} \to [0,\infty]$ such that there exists a constant $\kappa > 0$ (which need not be known) such that

$$\kappa^{-1} F(E) \leq \gamma_B(E) \leq \kappa\, F(E)$$

for all sets $E \subset \mathbb{C}$.

(6.4) The case $B = \text{lip}\,\alpha (0 < \alpha < 1)$ is simplest to describe. In this case γ_B is comparable to $M_*^{1+\alpha}$, <u>lower</u> $(1+\alpha)$ - <u>dimensional Hausdorff content</u> [19]. This content is defined as follows.

For $h : [0,\infty] \to [0,\infty]$ and $E \subset \mathbb{C}$ we define

$$M_h(E) = \inf_{S} \sum_{D \in S} h(\text{diam } D)$$

where S runs over all countable coverings of E by open discs. For instance, in case $h(r) = r^\beta$, M_h is denoted M^β and is called β-dimensional Hausdorff content or size infinity approximating β-dimensional Hausdorff measure. We define

$$M_*^\beta(E) = \sup_h M_h(E)$$

where h runs over all functions $[0,\infty] \to [0,\infty]$ such that $h(r) \leq r^\beta$ and $r^{-\beta}h(r) \to 0$ as $r \downarrow 0$.

It turns out [20] that $M_*^\beta(E) = M^\beta(E)$ if E is open, so Theorem (6.2) specializes to the following explicit result.

<u>Theorem</u>. <u>Let</u> $B = \text{lip}\,\alpha (0 < \alpha < 1)$. <u>Then</u> $A(X) = R(X)$ <u>if and only if there exists</u> $\kappa > 0$ <u>such that</u>

$$M^{1+\alpha}(D \sim X) \geq \kappa\, M_*^{1+\alpha}(D \sim \text{int } X)$$

for all open discs D.

Note that this is a metric condition. It implies that if $X = \phi(Y)$ where $\phi : \mathbb{C} \to \mathbb{C}$ is biLipschitzian, then $R(X) = A(X)$ if and only if $R(Y) = A(Y)$.

(6.5) In case $B = L^p$ $(2 \leq p < \infty)$, the capacity γ_B is of the potential theoretic kind, and has a couple of explicit descriptions [11]. One is that for compact E, $\gamma_B(E)$ is comparable to

$$\inf_u \|u\|_{W^{1,q}}$$

where u runs over all functions in $\mathcal{D}$ with $u \geq 1$ on E. Here q is the conjugate index to p, i.e. $\frac{1}{p} + \frac{1}{q} = 1$, and the $W^{1,q}$ norm of u is the L^q norm of $|u| + |\nabla u|$. Another description is that $\gamma_B(E)$ is comparable to

$$\sup_\mu \mu(E) \, ,$$

where μ runs over all positive measures supported on E such that the potential $K * \mu$ has L^p norm at most 1, where $K(z) = \frac{1}{|z|}$ is the Newtonian kernel. In the Hilbert space case, $p = 2$, which was the first to be cleared up [9], γ_B is the logarithmic capacity.

For all p, the capacity γ_B is a true Choquet capacity, and as a result it generates a corresponding fine topology on the plane. For those unfamiliar with such things, it may be helpful to describe a similar fine topology. The density topology on the plane is the topology for which a set N is a neighbourhood of a point a if $a \in N$ and $\mathbb{C} \sim N$ has area density 0 at a, i.e.

$$\lim_{r \downarrow 0} \frac{m\{B(a,r) \sim N\}}{\pi r^2} = 0 \, .$$

This topology on $\mathbb{C}$ is finer than the Euclidean topology, yet it is still connected and Baire (i.e. second category in itself). The fine topology associated to γ_B is finer than the density topology. It may be described as follows. A function $f : \mathbb{C} \to \mathbb{C}$ is said to be finely-continuous if given $\varepsilon > 0$ there exists an open set U, with $\gamma_B(U) < \varepsilon$, such that f is continuous (in the usual Euclidean sense) on $\mathbb{C} \sim U$. The fine topology generated by γ_B is the least topology such that each finely-continuous function is continuous, i.e. it is the topology with subbase $\{f^{-1}(V) : f$ is finely-continuous, V is a Euclidean open set$\}$. In case $B = L^2$, this fine topology is the standard fine topology of potential theory, namely the pullback topology generated by the superharmonic functions.

In terms of the fine topology of γ_B, $B = L^p$, the solution of main problem (2) may be expressed as follows.

Theorem. Let $2 \leq p < \infty$, $B = L^p$, and let X be compact in $\mathbb{C}$. Then $R(X) = A(X)$ if and only if the fine closure of $\mathbb{C} \sim X$ equals the set of fine accumulation points of $\mathbb{C} \sim \text{int } X$.

For $p = 2$, this is precisely the condition for each function, continuous on X and harmonic on $\text{int } X$, to be a uniform limit on X of functions harmonic on a neighbourhood of X. Thus L^2 analytic approximation is equivalent to L^∞ harmonic approximation. It would be interesting to see a more direct proof of this mysterious fact.

(6.6) The capacity γ_C corresponding to $B = C$ is the least well understood. It is known as continuous analytic capacity, and is usually denoted α. It was introduced by Dolženko. The associated outer capacity γ_C^*, defined by

$$\gamma_C^*(E) = \inf\{\gamma_C(U) : E \subset U \text{ open}\}$$

is the analytic capacity of Ahlfors. For any set E, $\gamma_c^*(E)$ is at most the logarithmic capacity of E, with equality for connected sets. For any set E, $\gamma_c(E)$ is at least the Newtonian capacity of E (defined using the kernel $|z|^{-1}$). In particular, $\gamma_c(E) \geq \{m(E)/\pi\}^{\frac{1}{2}}$. If $M^1(E) < \infty$, i.e. if E has finite outer length, then $\gamma_c(E) = 0$. If $M^\beta(E) > 0$ for some $\beta > 1$, then $\gamma_c(E) > 0$, so in terms of Hausdorff dimension the break-point for nullity of γ_c occurs at dimension 1. However, γ_c is not comparable to any M_h. A reference for the above facts if [6]. It is conjectured that γ_c^* is comparable to a one-dimensional Favard content (or integralgeometric content), calculable in terms of projections of E in almost every direction. See [12] for an account of progress on this, due to A. P. Calderon and S. Ya Havinson.

In principle, γ_c may be computed by a method due to P. Garabedian [22]. It suffices to calculate it for smoothly bounded compact sets with connected complement. Let E be a compact set with smooth boundary Γ, (possibly having several components) and with $\Omega = S^2 \sim E$ connected. Let $E^2(\Omega)$ denote the Smirnov space of all functions f, analytic in Ω, with non-tangential boundary values in $L^2(\Gamma, ds)$, where ds denotes arc length on Γ. Thus $E^2(\Omega)$ is a Hilbert space, with inner product

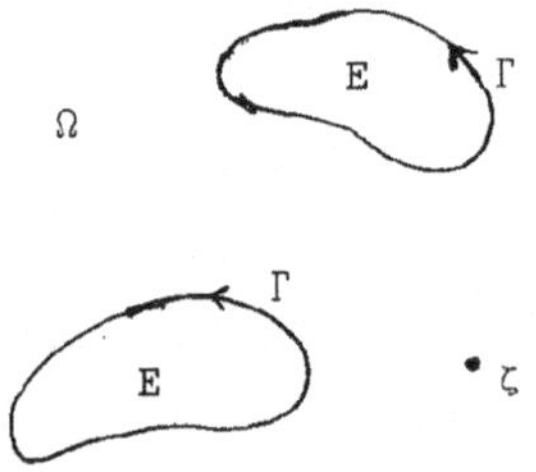

Figure 3

$$\langle f,g \rangle = \int_\Gamma f(z)\, \overline{g(z)}\, ds ,$$

and for $\zeta \in \Omega$, evaluation at ζ is a continuous linear functional on $E^2(\Omega)$. Thus there exists a function $z \to K(z,\zeta)$, belonging to

$E^2(\Omega)$, such that

$$f(\zeta) = \int_\Gamma f(z)\ \overline{K(z,\zeta)}\ ds$$

for all $f \in E^2(\Omega)$. The function $K(z,\zeta)$ is called the Szegö kernel. The formula of Garabedian is

$$K(\infty,\infty) = \frac{1}{2\pi\gamma_c(E)} .$$

The number $K(\infty,\infty)$ is the square of the norm of the functional $f \to f(\infty)$ on $E^2(\Omega)$. It can be computed to any desired degree of accuracy if enough terms of an orthonormal basis for $E^2(\Omega)$ are available. An orthonormal basis can be obtained by the Gram-Schmidt process from any independent set with dense span. If E has components $E_1, \ldots, E_k$ and a_j is chosen in the interior of $E_j (j=1,\ldots,k)$, then the set of functions $(z-a_j)^{-n}$ $(1 \le j \le k\ ,\ n \ge 0)$ has dense span in $E^2(\Omega)$. Thus $\gamma_c(E)$ can in principle be computed to any desired accuracy. A lemma of E. Smith [22] shows that the following algorithm works. Let $u_0 = 1, u_1, u_2, \ldots$ be an enumeration of the functions $(z-a_j)^{-n}$ $(1 \le j \le k, n \ge 0)$. Let $p_{ij} = \langle u_i , u_j \rangle$, and form the infinite matrix $P = (p_{ij})$ $(0 \le i < \infty , 0 \le j < \infty)$. Then $K(\infty,\infty)$ is the $(0,0)$ entry of P^{-1}. It is the limit of the numbers

$$\frac{\det(p_{ij} : 1 \le i,\ j \le n)}{\det(p_{ij} : 0 \le i,\ j \le n)} .$$

In practice, even if E is a simple kind of set with a small number of components, the number of computations involved in the algorithm is enormous, since the evaluation of each inner product involves a line integral. Thus it is prohibitively expensive to obtain more than two significant figures for $\gamma_c(E)$.

The main problem in the theory of uniform rational approximation is that of understanding γ_c. In particular, it is important to determine whether or not γ_c is <u>quasi-subadditive</u>, in the sense that

$$\gamma_c(E \cup F) \leq \kappa\{\gamma_c(E) + \gamma_c(F)\}$$

for some constant κ, independent of E and F. See [4].

§7. First main problem for the hairy class.

(7.1) The first main problem has been solved for $B = C$, and $B = \text{lip}\,\alpha(0 < \alpha < 1)$, but not for $B = L_p(2 \leq p < \infty)$.

For $\phi \in \mathcal{D}$, let $\text{spt}\phi$ denote the closed support of ϕ, the closure of $\{z \in \mathbb{C} : \phi(z) \neq 0\}$. Let $D(\phi)$ be the least closed disc containing $\text{spt}\phi$, and let $d(\phi)$ denote the diameter of $D(\phi)$. Let

$$||\phi||_* = ||\phi||_\infty + d(\phi) \cdot ||\nabla\phi||_\infty .$$

Theorem [24,21]. *Let* $B = C$ *or* $\text{lip}\,\alpha(0 < \alpha < 1)$. *Let* X *be a compact subset of* $\mathbb{C}$. *Let* $f \in B(\mathbb{C})$. *Then* $f \in R(X)$ *if and only if there exists* $\kappa > 0$ *such that*

$$\left| \int f\, \bar{\partial}\phi\, dm \right| \leq \kappa\, ||\phi||_*\, \gamma_B\, (D(\phi) \sim X)$$

for all $\phi \in \mathcal{D}$.

The condition given in this theorem for $f \in R(X)$ may be viewed as a kind of weak analyticity condition. If $\text{spt}\,\phi \subset \text{int}\,X$, then it reduces to

$$\int f\, \bar{\partial}\phi\, dm = 0 ,$$

so it says that as a *distribution* f satisfies $\bar{\partial}f = 0$ on $\text{int}\,X$. It is well-known that this forces f to be analytic on $\text{int}\,X$. At a boundary point a, the integral condition places a restriction on f which is more or less stringent depending on how *thin* $\mathbb{C} \sim X$ is at the point a, where thinness is measured in terms of γ_B.

(7.2) It cannot be said that this rational approximation theory is in a satisfactory state. Not only are there many open problems, major and minor, but, more importantly, there is no coherent method for deriving the known results. Widely varying tools are used to tackle different classes B. It ought to be possible to develop axiomatic frameworks, one for smooth B and one for hairy B, within which all the central results can be proved. It is my hope that this article will prompt others to seek such frameworks.

The following list of references contains only the most recent sources available to me. Where possible, I referred to a textbook or monograph rather than the original paper.

References

[1] T. Bagby, Quasi topologies and rational approximation, J. Functional Analysis, 10 (1972) 259-268.

[2] J. Brennan, Approximation in the mean by polynomials on non-Caratheodory domains, Ark. Mat. (1977).

[3] A. Browder, Lectures on Function Algebras, Benjamin, 1969.

[4] A. M. Davie, Analytic capacity and approximation problems, Transactions AMS 171 (1972) 409-444.

[5] T. W. Gamelin, Uniform Algebras, Prentice-Hall, 1969.

[6] J. Garnett, Analytic Capacity and Measure, LNM 297, Springer, 1972.

[7] P. M. Gauthier, Meromorphic uniform approximation on closed subsets of open Riemann surfaces, Approximation Theory and Functiona Analysis, J. B. Prolla (ed.), North-Holland, 1979.

[8] ________, Analytic approximation on closed subsets of open Riemann surfaces, to appear.

[9] V. P. Havin, Approximation in the mean by analytic functions, Doklady Akad. Nauk SSSR, 178 (1968) 1025-1028.

[10] L. I. Hedberg, Approximation in the mean by analytic functions, Transactions AMS 163 (1972) 157-171.

[11] ________, Non-linear potentials and approximation in the mean by analytic functions, Math. Z. 129 (1972) 299-319.

[12] D. Marshall, Removable sets for bounded analytic functions, Contemporary Problems in Complex and Linear Analysis, S. N. Nikolskü (ed.), Nauka, 1980.

[13] A. O'Farrell, Point derivations on an algebra of Lipschitz functions, Proceedings RIA 80A (1980) 23-39.

[14] ________, Localness of certain Banach modules, Indiana Univ. Math. J. 24 (1975) 1135-1141.

[15] ________, Annihilators of rational modules, J. Functional Analysis 19 (1975) 373-389.

[16] ________, Rational approximation in Lipschitz norms I, Proceedings RIA 77A (1977) 113-115.

[17] ________, Rational approximation in Lipschitz norms II, Proceedings RIA 79A (1979) 103-114.

[18] ________, Lip 1 rational approximation, Journal LMS (2) 11 (1975) 159-164.

[19] ________, Hausdorff content and rational approximation in fractional Lipschitz norms, Transactions AMS 228 (1977) 187-206.

[20] ________, Continuity properties of Hausdorff content, Journal LMS (2) 13 (1976) 403-410.

[21] ________, Estimates for capacities and approximation in Lipschitz norms, J.f.d. Reine u. Angew. Math. 311/312 (1979) 101-115.

[22] E. P. Smith, The Garabedian function of an arbitrary compact set, Pacific J. Math. 51 (1974), 289-300.

[23] E. M. Stein, Singular Integrals and Differentiability Properties of Functions, Princeton 1970.

[24] A. G. Vitushkin, Analytic capacity of sets and problems of approximation theory, Russian Math. Surveys 22 (1967) 139-200.

SOME APPLICATIONS OF THE METRIC ENTROPY CONDITION TO HARMONIC ANALYSIS

Gilles Pisier[1]

Introduction:

It has been known for some time (cf. [3],[4]) that the a.s. continuity of Gaussian processes can be studied efficiently by considering an "entropy condition" relative to the index set equipped with a certain pseudo-metric associated to the random process. These methods also apply to processes which are not necessarily Gaussian, but which behave similarly, like for instance Bernoulli series. In this context, these techniques have already been applied successfully to random Fourier series (see Chapters 3, 4, 5 of [7]) and to Harmonic Analysis (see Chapter 6 of [7]). The object of this paper is to present some more applications, but in a non-Gaussian (not even close to Gaussian) setting.

Only recently (cf. [6],[10]), it became clear that Dudley's theorem (ensuring that the metric entropy condition is a sufficient condition for the a.s. continuity of a Gaussian process) can be extended to a very general setting. The extension is as follows. Consider a probability space $(\Omega,\mathcal{A},\mathbb{P})$, a convex increasing function $\psi\colon R_+ \to R_+$ with $\psi(0) = 0$, and the Orlicz space $L^{\psi}(\Omega,\mathcal{A},\mathbb{P})$ (for more precisions see below). Let T be a compact metric space and let

$$t \to X_t \in L^{\psi}(\Omega,\mathcal{A},\mathbb{P})$$

be a continuous map from T into L^{ψ}. We may equip T with a pseudo-metric associated to X:

---

(1) These notes are based on lectures held at the University of Connecticut (Storrs) in October 1980 and at Collège de France (Paris) in January 1981.

$$\forall t,s \in T \quad d_X(s,t) = \|X_t - X_s\|_\psi .$$

We denote by $N(T,d_X;\varepsilon)$ the smallest number of open balls of radius ε for the metric d_X which form a covering of T. Then, the metric entropy condition

$$(0.1) \qquad \int_0 \psi^{-1}\big(N(T,d_X;\varepsilon)\big)\, d\varepsilon < \infty$$

is sufficient for the a.s. continuity of the sample paths of a version of the random process $(X_t)_{t\in T}$. Dudley's theorem corresponds to the case $\psi(x) = \exp x^2 - 1$. For the proof of this theorem, we follow closely [10] which treats the case $\psi(x) = x^p$; as pointed by Fernique, only minor modifications of the arguments of [10] are necessary to obtain the theorem in the above generality.

The preceding result generalizes a classical theorem of Kolmogorov; it is also related to the more recent work of Garsia [5].

The above theorem can be applied to Harmonic Analysis in the following manner: Let G be a compact group, and let f be an element of a translation invariant Banach space X of functions on G. Let

$$d_f^X(s,t) = \|f_t - f_s\|_X \quad \text{for} \quad t,s \in G$$

and

$$f_t(x) = f(x+t) .$$

We assume that the "trigometric polynomials" are dense in X. Then, the condition

$$(0.2) \qquad \int_0 \psi^{-1}\big(N(G,d_f^X;\varepsilon)\big)\, d\varepsilon < \infty$$

implies that f can be written as

$$f = \sum_{n=1}^{\infty} h_n * k_n$$

(0.3)

$$\text{with} \quad \sum_1^{\infty} \|h_n\|_X \|k_n\|_\phi < \infty$$

where ϕ is the Orlicz function conjugate to ψ (see Theorem 2.1). In the particular case $X = L^\phi = L^\psi = L^2$, we obtain an extension of S. Bernstein's classical theorem on absolutely convergent Fourier series to an "abstract" setting.

But the real interest of the preceding result is that in some cases (precisely if $X = L^2$ and $\psi(x) = \exp x^q - 1$ for $2 \leq q < \infty$), its converse is also true, that is to say (0.3) implies (0.2). This is the content of Theorem 2.6 below which improves the results of a preceding paper (cf. [11] § 4). We obtain also an interpolation theorem (using the Lions-Peetre method) for the Banach spaces of functions of the form (0.3) with $X = L^2$ and $\psi(x) = \exp x^q - 1$ with $2 \leq q < \infty$ (see Corollary 2.12 below).

Finally, a word on "self-completeness": we have written the first section (about the continuity of stochastic processes) with an analyst in mind as a reader; therefore, this section is very detailed and practically self-contained. On the other hand, section 2 depends on several results which are stated without proof.

§1. The entropy condition

We first recall some notations. Let (Ω, Σ, μ) be a measure space and let

$$\psi : R_+ \to R_+$$

be an increasing convex function such that $\psi(0) = 0$. The Orlicz space $L^\psi(\Omega, \Sigma, \mu)$ is the space of all measurable functions Z in $L^0(\Omega, \Sigma, \mu)$ such that there is a $c > 0$ for which

$$\int \psi(|\tfrac{Z}{c}|)\, d\mu < \infty .$$

As usual, for any such Z we define

$$\|Z\|_\psi = \inf \{c > 0 \mid \int \psi(|\tfrac{Z}{c}|)\, d\mu \leq 1\} .$$

Equipped with this norm, the space $L^\psi(\Omega,\Sigma,\mu)$ is a Banach space.

Now let (T,d) be a compact pseudo-metric space. We will denote $N(T,d;\varepsilon)$ the smallest number of open balls of radius ε (for the pseudo-metric d) which form a covering of T. We will denote by $|A|$ the cardinality of a set A. If S is a subset of a pseudo-metric space (T,d), we will denote by $\mathrm{diam}(S)$ the diameter of S, i.e.

$$\mathrm{diam}(S) = \sup \{d(t,s) \mid t,s \in S\} .$$

The main result of this section is the following.

<u>Theorem 1.1</u>: Let ψ be as above. Assume that

$$\int_0 \psi^{-1}(N(T,d;\varepsilon))\, d\varepsilon < \infty$$

Then, every random process $\{X_t \mid t\in T\} \subset L^\psi(\Omega,a,\mathbb{P})$ (where $(\Omega,\mathcal{a},\mathbb{P})$ is an arbitrary probability space) such that:

(1.1) $$\forall t,s \in T \quad \|X_t - X_s\|_\psi \leq d(t,s)$$

admits a version with continuous sample paths and verifies the inequality:

(1.2) $$\mathbb{E} \sup_{t,s\in T} |X_t - X_s| \leq K \int_0^D \psi^{-1}(N(T,d;\varepsilon))\, d\varepsilon$$

where K is an absolute constant (independent, in particular, of T) and where

$$D \leq \sup \{d(s,t) \mid s,t \in T\}$$

is defined as the smallest $\varepsilon > o$ such that $N(T,d;\varepsilon) = 1$.

Remark: For the proof of this theorem, we follow essentially [10]; I have included a slight improvement over [10] which was kindly pointed out to me by X. Fernique. Moreover, I should mention that N. Kôno [6] proved a result which is very close to the above; at the time of [10], I was not aware of Kôno's paper [6].

Remark 1.2: In the preceding statement (and throughout this paper) the variable

$$\sup_{t,s\in T} |X_t - X_s|$$

should be interpreted as the lattice-supremum of the variables $\{X_t - X_s|\ t,s\in T\}$ in the Banach lattice $L^1(\Omega,\mathcal{A},\mathbb{P})$. In other words, the quantity $\mathbb{E} \sup_{t,s\in T} |X_t - X_s|$ should be interpreted as:

$$\sup \{\mathbb{E} \sup_{(t,s)\in F} |X_t - X_s| \ \Big|\ F \subset T\times T,\ F \text{ finite}\} .$$

As a consequence, it follows that, to prove (1.2), we may as well assume that T is a finite set.

Remark 1.3: Let $\eta = \sup \{\|X_t - X_s\|_\psi \ | t,s\in T\}$ and let $d'(t,s) = \|X_t - X_s\|_\psi$; if we apply (1.2) to the same process, but to the metric d' instead of d, we clearly obtain:

$$\mathbb{E} \sup_{t,s\in T} |X_t - X_s| \leq K \int_0^\eta \psi^{-1}\big(N(T,d;\varepsilon)\big)\, d\varepsilon . \tag{1.3}$$

First part of the proof of Theorem 1.1: We wish to first show that everything reduces to the proof of the inequality (1.2). Indeed, suppose that we already know that all the processes verifying (1.1) must verify (1.2); we claim that any process $(X_t)_{t\in T}$ verifying (1.1) must then have a version with continuous sample paths. To prove this claim, we will need the following elementary lemmas:

Lemma 1.4: Let $\{Z_1,\dots,Z_n\} \subset L^0(\Omega,\mathcal{a},\mathbb{P})$ be a finite set of random variables. For any $\varepsilon_o > 0$ there exists a σ-subalgebra $\mathcal{B} \subset \mathcal{a}$ which is generated by a countable partition of Ω and such that:

$$\sup_{i \le N} \text{ ess sup } |Z_i - \mathbb{E}^{\mathcal{B}} Z_i| \le \varepsilon_o .$$

Proof: Given $(m_1,\dots,m_N) \in \mathbb{Z}^N$, define the set

$$\Omega(m_1,\dots,m_N) = \bigcap_{i \le N} \{m_i\ \varepsilon_o \le Z_i < (m_i+1)\varepsilon_o\} .$$

Then the partition $\{\Omega(m_1,\dots,m_N) \mid (m_1,\dots,m_N) \in \mathbb{Z}^N\}$ generates a σ-algebra $\mathcal{B}$ verifying the required property. q.e.d.

Since $\{X_t | t \in T|$ is a relatively compact subset of $L^\psi(\Omega,\mathcal{a},\mathbb{P})$, we immediately deduce from Lemma 1.4:

Lemma 1.5: For any $\varepsilon_o > 0$, we can find a σ-algebra $\mathcal{B} \subset \mathcal{a}$ as above and such that

$$\sup_{t \in T} \|X_t - E^{\mathcal{B}} X_t\|_\psi < 3\varepsilon_o .$$

Proof: By assumption, $N(T,d;\varepsilon) < \infty$ for any $\varepsilon > o$ and (X_t) verifies (1.1). Therefore, there exists a finite set $\{Z_1,\dots,Z_N\}$ in $L^\psi(\Omega,\mathcal{a},\mathbb{P})$ such that for each t in T $\exists\ i \le N$ so that

$$\|X_t - Z_i\|_\psi < \varepsilon .$$

Let then $\mathcal{B}$ be given by Lemma 1.4, and we have:

$$\|X_t - \mathbb{E}^{\mathcal{B}} X_t\|_\psi \le \|X_t - Z_i\|_\psi + \|Z_i - \mathbb{E}^{\mathcal{B}} Z_i\|_\psi + \|\mathbb{E}^{\mathcal{B}} (Z_i - X_t)\|_\psi$$

$$\le 3\varepsilon_o . \qquad \text{q.e.d.}$$

Finally, we observe that if $\mathcal{B} \subset \mathcal{A}$ is as above, then the paths of the process $(\mathbb{E}^{\mathcal{B}} X_t)_{t \in T}$ are a.s. continuous. Indeed, let A be an atom of $\mathcal{B}$ with $\mathbb{P}(A) > 0$, then, on A, we have

$$\mathbb{E}^{\mathcal{B}} X_t = \frac{1}{\mathbb{P}(A)} \int_A X_t \, d\mathbb{P} = \mathbb{P}(A)^{-1} < X_t, 1_A > .$$

and the map $t \to \langle X_t, 1_A \rangle$ is continuous since 1_A is a continuous linear form on the space $L^{\psi}(\Omega, \mathcal{A}, \mathbb{P})$ and $t \to X_t$ is a continuous function from T into L^{ψ}.

We can now prove the announced claim: Since

$$\int_0 \psi^{-1}\big(N(T,d;\varepsilon)\big) \, d\varepsilon < \infty ,$$

we can find a sequence $(\eta_n)_{n=1}^{\infty}$, $\eta_n > 0$, such that

$$\sum_{n=1}^{\infty} \int_0^{6\eta_n} \psi^{-1}\big(N(T,d;\varepsilon)\big) \, d\varepsilon < \infty . \tag{1.4}$$

By the two lemmas above, we can find a sequence $\mathcal{B}_n \subset \mathcal{A}$ of σ-algebras as above such that:

$$\sup_{t \in T} \| X_t - \mathbb{E}^{\mathcal{B}_n} X_t \|_{\psi} \leq 3\eta_n ;$$

let

$$Y_t^n = X_t - \mathbb{E}^{\mathcal{B}_n} X_t .$$

Obviously, we have

$$\| Y_t^n - Y_s^n \|_{\psi} \leq 2 \| X_t - X_s \|_{\psi} \leq 2d(t,s) ,$$

and hence by remark 1.3 (and assuming (1.2)):

$$\mathbb{E} \sup_{t,s} | Y_t^n - Y_s^n | \leq K \int_0^{6\eta_n} \psi^{-1}\big(N(T,d;\varepsilon/2)\big) \, d\varepsilon . \tag{1.5}$$

So that, if we set $Y_t^0 \equiv 0$ and $Z_t^n = Y_t^n - Y_t^{n-1}$, we find by (1.4) that

$$\text{(1.6)} \qquad \mathbb{E} \sum_{n=1}^{\infty} \sup_{t,s\in T} |Z_t^n - Z_s^n| < \infty .$$

By the observation following lemma 1.5, the process $(Z_t^n)_{t\in T}$ has continuous sample paths; therefore, if we define $\tilde{X}_t$ as the sum of the series $\sum_{n=1}^{\infty} Z_t^n$, the estimate (1.6) shows that the sample paths of $(\tilde{X}_t)$ are a.s. continuous. In conclusion, since it is obvious that $\tilde{X}_t = X_t$ a.s. for each t in T, the process $(\tilde{X}_t)_{t\in T}$ is a version of $(X_t)_{t\in T}$ and this ends the proof of Theorem 1.1, modulo the proof of (1.2).

For the proof of (1.2), we will need the following easy Lemma:

Lemma 1.6: Let $\{Z_1,\ldots,Z_N\} \subset L^{\psi}(\Omega,\mathcal{A},\mathbb{P})$, then

$$\mathbb{E} \sup_{i\le N} |Z_i| \le \psi^{-1}(N) \sup_{i\le N} \|Z_i\|_{\psi} .$$

Proof: By homogeneity, we may assume $\sup_{i\le N} \|Z_i\|_{\psi} \le 1$, or equivalently $\mathbb{E}\,\psi(Z_i) \le 1$ for $i = 1,2,\ldots,N$. Since ψ is convex and increasing, we have

$$\psi(\mathbb{E} \sup_{i\le N} |Z_i|) \le \mathbb{E}\,\psi(\sup_{i\le N} |Z_i|) = \mathbb{E} \sup_{i\le N} \psi(|Z_i|)$$

$$\le \mathbb{E} \sum_{i=1}^{N} \psi(|Z_i|) \le N ,$$

and therefore

$$\mathbb{E} \sup_{i\le N} |Z_i| \le \psi^{-1}(N) \qquad \text{q.e.d.}$$

The basic principle of the proof of (1.2) is then to decompose the process $(X_t)_{t\in T}$ as a series of simpler processes to which we will apply the preceding lemma.

Proof of (1.2): The beginning of the proof is entirely classical. We denote

$$\delta_n = D4^{-n}, \quad n = 0,1,2,\ldots$$

and

$$N_n = N(T,d;\delta_n).$$

We denote $B(t,\delta)$ the open ball with center t and radius δ. We know that there are points $(t_j^n)_{j\leq N_n}$ in T such that

$$T = \bigcup_{j\leq N_n} B(t_j^n, \delta_n).$$

Therefore, we can clearly find a partition $(A_j^n)_{j\leq N_n}$ of T such that

(1.7)
$$A_j^n \subset B(t_j^n, \delta_n)$$

and, in particular, the diameter of A_j^n is less than $2\delta_n$. We then define the following approximating process:

$$X_t^n(\omega) = \sum_{j\leq N_n} 1_{A_j^n}(t)\, X_{t_j^n}(\omega).$$

We have

$$X_t = X_t^0 + \sum_{n\geq 1} X_t^n - X_t^{n-1}.$$

We observe that X_t^0 does not depend on t; therefore,

(1.8)
$$\sup_{t,s\in T} |X_t - X_s| \leq 2 \sum_{n\geq 1} \sup_t |X_t^n - X_t^{n-1}|$$

$$\leq 2 \sum_{n\geq 1} \sup_{(i,j)\in\Lambda_n} |X_{t_j^n}^n - X_{t_i^n}^n|,$$

where Λ_n is the subset of $\{1,2,\ldots,N_n\}\times\{1,\ldots,N_{n-1}\}$ formed by the couples (i,j) such that

$$A_j^n \cap A_i^{n-1} \neq \emptyset .$$

Therefore, (1.8) and Lemma 1.6 give us:

(1.9) $$\mathbb{E} \sup_{t,s\in T} |X_t - X_s| \leq 2 \sum_{n\geq 1} \psi^{-1}(|\Lambda_n|) \sup_{(i,j)\in\Lambda_n} \|X_{t_j^n} - X_{t_i^n}\|_\psi .$$

Now, if $(i,j) \in \Lambda_n$ and if $t \in A_j^n \cap A_i^{n-1}$, we have:

$$\|X_{t_j^n} - X_{t_i^{n-1}}\|_\psi \leq \|X_{t_j^n} - X_t\|_\psi + \|X_t - X_{t_i^{n-1}}\|_\psi$$

$$\leq \text{diam}\,(A_j^n) + \text{diam}\,(A_i^{n-1}) ,$$

and hence we finally deduce from (1.9):

(1.10) $$\mathbb{E} \sup_{t,s\in T} |X_t - X_s| \leq 2 \sum_{n\geq 1} \psi^{-1}(|\Lambda_n|) \sup_{i,j} \text{diam}\,(A_j^n) + \text{diam}\,(A_i^{n-1}) .$$

In the particular case $\psi(x) = \exp x^2 - 1$ (or even $\psi(x) = \exp x^\alpha - 1$ for $0 < \alpha < \infty$), the proof is then easily completed: indeed, the trivial estimate $|\Lambda_n| \leq N_n N_{n-1} \leq N_n^2$ suffices because $\psi^{-1}(N_n^2)$ and $\psi^{-1}(N_n)$ are essentially equivalent; (note that, in this particular case, the reduction justified in the first part of the proof is not needed). However, in the general case we have to work a little more; we will need to prove the following lemma.

Lemma 1.7: Let Δ_1, Δ_2 be two partitions of T. We will say that $\Delta_1 \succ \Delta_2$ if every element of Δ_2 is included in a (necessarily unique) element of Δ_1. Let (T,d) be a finite pseudo-metric space. Then, there exist partitions

$$\Delta_1 \succ \Delta_2 \succ \ldots\ldots \succ \Delta_n \succ \ldots$$

such that

$$|\Delta_n| < N_n$$

and

$$\sup \{\text{diam}\,(A) \quad | A \in \Delta_n\} \leq 4\,\delta_n \qquad \forall n \geq 1 .$$

With the help of this lemma (which is proved below), we can now complete the proof of (1.2): Indeed, we may assume (cf. remark 1.2) that T is a finite set, so that we may apply the preceding lemma. We will use the partitions Δ_n instead of $(A_j^n)_{j\leq N_n}$. We may repeat the argument which leads to (1.10), but since $\Delta_{n-1} \succ \Delta_n$, we have the advantage that

$$|\{(A,B) \in \Delta_n \times \Delta_{n-1} \mid A \cap B \neq \emptyset\}| \leq |\Delta_n| \leq N_n .$$

Therefore, (1.10) (with Δ_n replacing $(A_j^n)_{j\leq N_n}$, for $n = 1,2,\ldots$) becomes:

$$\mathbb{E} \sup_{t,s\in T} |X_t - X_s| \leq 2 \sum_{n\geq 1} \psi^{-1}(N_n) \sup_{\substack{A\in\Delta_n \\ B\in\Delta_{n-1}}} \big(\operatorname{diam}(A) + \operatorname{diam}(B)\big)$$

$$\leq 16 \sum_{n\geq 1} \psi^{-1}(N_n)\, \delta_{n-1} .$$

Finally, comparing the preceding series with the corresponding integral, we obtain

$$\mathbb{E} \sup_{t,s\in T} |X_t - X_s| \leq 64 \int_0^D \psi^{-1}\big(N(T,d;\varepsilon)\big)\, d\varepsilon .$$

q.e.d.

<u>Proof of Lemma 1.7</u>: Let Δ be a partition of T. We set $\pi(\Delta) = \sup_{A\in\Delta} \operatorname{diam}(A)$. Denote Δ_n' the partition formed by the sets $(A_j^n)_{j\leq N_n}$ introduced above. We have $\forall n = 1,2,\ldots$ $|\Delta_n'| \leq N_n$ and $\pi(\Delta_n') \leq 2\delta_n$, but we do <u>not</u> know a priori that Δ_{n+1}' is a refinement of Δ_n'. To obtain this additional property, we will have to modify the partitions (Δ_n'), using a relatively simple procedure: Since T is finite, we know a priori that there exists an integer N such that, for $n \geq N$, Δ_n' is the trivial partition of T consisting of singletons. Therefore, we have $\Delta_n' \succ \Delta_{n+1}'$ for all $n \geq N$. We define

$\Delta_n = \Delta'_n$ for all $n \geq N$. For $n < N$, we will define $\Delta_n > \Delta_{n+1} > \ldots . > \Delta_N \ldots .$ by induction on n. For each $n \leq N$, we denote

$$\lambda_n = 2^{-n} \sum_{j=n}^{j=N} 2^{j+1} \delta_j .$$

Now assume that we have found partitions

$$\Delta_n \prec \Delta_{n+1} \prec \ldots . \prec \Delta_N \prec \ldots .$$

such that

$$|\Delta_j| \leq N_j \quad \text{and} \quad \pi(\Delta_j) \leq \lambda_j \quad \text{for all} \quad j \geq n .$$

We can then construct Δ_{n-1} as follows. Let $A \subset T$ be a set belonging to Δ'_{n-1}; we may associate to it the set

$$[A] = \bigcup \{B \mid B \in \Delta_n , \quad B \cap A \neq \emptyset\} .$$

The collection

$$\mathcal{C} = \{[A], \ A \in \Delta'_{n-1}\}$$

is then a covering of T with sets belonging to the Boolean algebra generated by the partition Δ_n. We have clearly $|\mathcal{C}| \leq |\Delta'_{n-1}| \leq N_{n-1}$, and $\forall C \in \mathcal{C}$

$$\operatorname{diam}(C) \leq 2\ \delta_{n-1} + 2\ \lambda_n = \lambda_{n-1} .$$

Obviously, by replacing each C in $\mathcal{C}$ by a suitable subset of C, we can now form a partition Δ_{n-1} of T such that $\Delta_{n-1} \succ \Delta_n$, $|\Delta_{n-1}| \leq N_{n-1}$ and $\pi(\Delta_{n-1}) \leq \lambda_{n-1}$. By induction, this proves the announced result, since we have

$$\lambda_n = 2^{-n} \sum_{j=n}^{j=N} 2^{j+1} \delta_j = D\ 2^{-n} \sum_{j=n}^{j=N} 2^{-j+1}$$

$$\leq 4\ D\ 2^{-2n} = 4\ \delta_n .$$

This concludes the proof of Lemma 1.7.

The main interest of Theorem 1.1 is that, for certain particular functions, the converse is also true. For instance, Fernique [4] proved (essentially) the following result:

Theorem 1.8: Assume that $T = G$ is a compact group, and that d is a continuous pseudo-metric on G which is invariant under left translations, and assume moreover that (G,d) is a Hilbertian pseudo-metric space [i.e. there exists a Hilbert space H and a map $\omega: G \to H$ such that $d(s,t) = \|\omega(s) - \omega(t)\|$ for all t,s in G]. Denote

$$\psi_2(x) = \exp|x|^2 - 1 .$$

Then the following conditions are equivalent:

i) Any process $(X_t)_{t\in G}$ such that

$$\forall t,s\in G \qquad \|X_t - X_s\|_{\psi_2} \leq d(t,s)$$

admits a version with continuous sample paths.

ii) The (necessarily unique) Gaussian process $(X_t)_{t\in G}$, which verifies $\|X_t - X_s\|_2 = d(s,t)$ for all t,s in G and $X_o = 0$, admits a version with continuous sample paths.

iii)
$$\int_0 \psi_2^{-1} \big(N(G,d;\varepsilon)\big)\, d\varepsilon < \infty .$$

(Note that iii) is equivalent to $\int_0 \big(\mathrm{Log}\, N(G,d;\varepsilon)\big)^{\frac{1}{2}}\, d\varepsilon < \infty$).

In view of the last result, it is natural to raise the following:

Problem 1.9: For which functions ψ other than ψ_2 is the equivalence i) $\Leftrightarrow$ iii), in the preceding statement, still valid.

In [10], it is pointed that Fernique's proof can be extended to the case of functions of the form $\psi_q(x) = \exp|x|^q - 1$ for $0 < q \leq 2$. Of course, the group structure is essential in all these results.

The case $\psi(x) = x^2$ which seems particularly interesting is entirely open. We only could prove the following simple result (which does not use the group structure):

Proposition 1.10: Let (T,d) be a compact pseudo-metric space such that any process $(X_t)_{t\in T}$ verifying

$$(1.11) \qquad \forall t,s \in T \qquad \|X_t - X_s\| \leq d(t,s)$$

has a version with continuous sample paths. Then

$$\sup_{0<\varepsilon<D} \varepsilon\left(N(T,d;\varepsilon)\right)^{\frac{1}{2}} < \infty .$$

Proof: By a routine argument (for more details on this point, the reader may consult p. 8 in [10]), one can show the above assumption implies "automatically" the existence of a constant C such that any process $(X_t)_{t\in T}$ verifying (1.11) must also satisfy

$$(\mathbb{E} \sup_{t,s\in T} |X_t - X_s|^2)^{\frac{1}{2}} \leq C .$$

Now, fix $\varepsilon > 0$ and let S be a maximal subset of T such that $\forall t,s\in S$, $t \neq s$, $d(t,s) \geq \varepsilon$. We clearly have

$$|S| \geq N(T,d;\varepsilon) .$$

Let $(A_t)_{t\in S}$ be a collection of disjoint measurable sets on some probability space $(\Omega,\mathcal{A},\mathbb{P})$ such that $\mathbb{P}(A_t) = |S|^{-1}$ for all t in S. We define

$$\forall s\in S \qquad Y_s = \varepsilon 1_{A_s} |S|^{\frac{1}{2}} 2^{-\frac{1}{2}} .$$

We obviously have

$$\|Y_t - Y_s\|_2 \leq d(t,s) \quad \forall t,s \in S .$$

Therefore, the map $s \to Y_s$ is a Lipschitzian function with constant 1 from S T into $L^2(\Omega,\mathcal{a},\mathbb{P})$. By a well known result (cf. [12], Theorem 11.3), there exists an extension, which we again denote $(Y_t)_{t\in T}$, such that

$$\|Y_t - Y_s\|_2 \leq d(t,s) \quad \text{for all} \quad t,s \quad \text{in} \quad T .$$

By the first part of the proof, we must have

$$(\mathbb{E} \sup_{t,s\in T} |Y_t - Y_s|^2)^{\frac{1}{2}} \leq C .$$

But, since

$$|Y_t - Y_s| = \varepsilon |S|^{\frac{1}{2}} 2^{-\frac{1}{2}} (1_{A_t} + 1_{A_s})$$

for $t \neq s$ in S, we have (if $|S|>1$):

$$\sup_{t,s\in S} |Y_t - Y_s| \geq 2^{-\frac{1}{2}} \varepsilon |S|^{\frac{1}{2}} \geq 2^{-\frac{1}{2}} \varepsilon \big(N(T,d;\varepsilon)\big)^{\frac{1}{2}} ;$$

hence, we must have

$$\varepsilon \big(N(T,d;\varepsilon)\big)^{\frac{1}{2}} \leq \sqrt{2}\, C$$

so that finally

$$\sup_{0<\varepsilon<D} \varepsilon \big(N(T,d;\varepsilon)\big)^{\frac{1}{2}} \leq \sqrt{2}\, C . \quad \text{q.e.d.}$$

<u>Remark 1.11</u>: The preceding proof of Theorem 1.1 works also in the case of a Banach space valued process $\{X_t \mid t \in T\} \subset L^{\psi}(\Omega,\mathcal{a},\mathbb{P};B)$ where B is a Banach space. The modifications of the proof are minor. Also, we should point out that in the case $\psi(x) = x^p$ and $p > 1$, Theorem

1.1 applies also to processes $\{X_t | t \in T\}$ which are not necessarily in L^p but only in weak L^p. For details on this point, the interested reader is referred to [10].

§2. Applications to Harmonic Analysis

We now give some applications to Harmonic Analysis and in particular to the preduals of certain spaces of multipliers.

We first need to introduce some notations: Let G be a compact Abelian group with its normalized Haar measure m. Denote $T(G)$ the linear space of all "trigonometric" polynomials on G, and $\hat{G}$ the dual group. We consider a norm $\| \ \|_X$ on $T(G)$ which is invariant under translation, and we denote X the Banach space which is the completion of $T(G)$ for this norm; we again denote $\| \ \|_X$ the norm after completion.

We will say that a function $\phi : R_+ \to R_+$ is a Young function if it is convex, increasing and $\phi(0) = 0$. We will say that a couple (ϕ,ψ) of Young functions are "in duality" if every f in $L^\psi(G)$ verifies

$$\int |fg| \, dm < \infty \quad \text{for all } g \text{ in } L^\phi(G)$$

and if, moreover, we can identify (with equivalent norms) the space $L^\psi(G)$ with the dual of $L^\phi(G)$, by the usual duality:

$$\forall g \in L^\phi(G) \quad \forall f \in L^\psi(G) \quad \langle f,g\rangle = \int fg \, dm .$$

We will denote $A(X,L^\phi)$ the space of all f in X which can be represented as

$$f = \sum_{n=1}^{\infty} h_n * k_n$$

with $h_n \in X$, $k_n \in L^\phi$ and $\sum_1^\infty \|h_n\|_X \|k_n\|_\phi < \infty$; we define

$$\|f\|_{A(X,L^\phi)} = \inf \{ \sum^\infty \|h_n\|_X \|k_n\|_\phi \}$$

where the infimum runs over all such representations. Equipped with this norm, the space $A(X,L^{\phi})$ becomes a Banach space.

Moreover, we will denote $M(X,L^{\psi})$ the space of all bounded operators $T : X \to L^{\psi}(G)$ which commute with the translations by elements of G. Such an operator T is necessarily of the form:

$$\forall f \in T(G) \qquad T(f) = \sum_{\gamma \in \hat{G}} \hat{T}(\gamma) \hat{f}(\gamma)\gamma ,$$

and the coefficients $(\hat{T}(\gamma))_{\gamma \in \hat{G}}$ uniquely determine the operator T. Equipped with the norm induced by the natural norm in the space of all bounded operators from X into L^{ψ}, this space becomes a Banach space.

It follows from a standard construction in harmonic analysis that the space $M(X,L^{\psi})$ can be identified with the dual of $A(X,L^{\phi})$ (where (ϕ,ψ) are in duality as explained above). The duality between $M(X,L^{\psi})$ and $A(X,L^{\phi})$ is defined by:

$$\forall T \in M(X,L^{\psi}) \qquad \forall f \in A(X,L^{\phi}) ,$$

$$\langle T,f \rangle = T(f)(0) .$$

Indeed, it is not hard to see that $T(f)$ has to be a continuous function on G so that $T(f)(0)$ makes sense. If $f \in T(G)$, we can write simply

$$\langle T,f \rangle = T(f)(0) = \sum_{\gamma \in \hat{G}} \hat{T}(\gamma) \hat{f}(\gamma) .$$

Note that $T(G)$ is dense in $A(X,L^{\phi})$. For f in X, we denote f_t the translated function, i.e. $f_t(x) = f(x+t)$. The theorem 1.1 of the preceding section has the following immediate application:

Theorem 2.1: Let X, ϕ, ψ be as above. Consider f in X and define $\forall t,s \in G \quad d_f^X(s,t) = \|f_s - f_t\|_X$. With these notations (and those of §1), if

(2.1) $$\int_0 \psi^{-1}(N(G,d_f^X;\varepsilon))\, d\varepsilon < \infty$$

then $f \in A(X,L^\phi)$ and we have:

(2.2) $$\|f\|_{A(X,L^\phi)} \leq C\{\,|\hat{f}(o)| + \int_0^D \psi^{-1}(N(G,d_f^X;\varepsilon))\, d\varepsilon\,\}$$

where C is a constant (depending only on ϕ,ψ and X) and where D is defined as before by

(2.3) $$D = \inf\{\,\varepsilon > 0 \mid N(G,d_f^X;\varepsilon) = 1\}.$$

Proof: Assume that f verifies (2.1). Consider T in the unit bal of $M(X,L^\psi)$. We have

$$\forall t,s \in G \quad \|T(f_t - f_s)\|_\psi \leq \|f_t - f_s\|_X .$$

Therefore, we may apply Theorem 1.1 to the random process $(T(f_t))_{t\in G}$ (considered as a random process on the probability space (G,m)). This shows that the process $t \to T(f_t)$ has a version with continuous sample paths. Equivalently, of course, this means simply that the function $T(f)$ is in $C(G)$. From (1.2), we obtain:

(2.4) $$\int \sup_{t,s\in G} |T(f_t)(\omega) - T(f_s)(\omega)|\, dm(\omega) \leq K\, I$$

with $I = \int_0^D \psi^{-1}(N(G,d_f^X;\varepsilon))\, d\varepsilon$ and with D as defined in (2.3). We have $T(f_t)(\omega) = T(f)(t+\omega)$, so that

$$\sup_{t,s} |T(f_t)(\omega) - T(f_s)(\omega)| \equiv \sup_{t,s} |T(f)(t) - T(f)(s)|$$

Hence, we deduce from (2.4)

$$\sup_{t,s} |T(f)(t) - T(f)(s)| \leq K\, I ,$$

and a fortiori

$$(2.5)\qquad |T(f)(0)| \leq \sup_t |T(f)(t) - \int T(f)(s)\, dm\,(s)| + |\hat{T}(0)\,\hat{f}(0)|$$

$$\leq K\, I + |\hat{f}(0)|\; |\hat{T}(0)|\;.$$

On the other hand, we have obviously

$$\|T(1)\|_{\psi} \leq \|1\|_{X}$$

(where 1 denotes here the function which is constantly equal to 1 on G). Hence, since $T(1) = \hat{T}(o)1$, $|\hat{T}(o)| \leq \|1\|_X(\|1\|_{\psi})^{-1}$. There fore, we deduce easily from (2.5) that

$$|T(f)(0)| \leq C'\,\{I + |\hat{f}(0)|\}$$

where C' depends only on X and ψ. Equivalently, we have

$$(2.6)\qquad \|f\|_{M(X,L^{\psi})^*} \leq C'\,\{I + |\hat{f}(0)|\}\,.$$

Since the norm in $M(X,L^{\psi})^*$ is equivalent on $T(G)$ with the norm in $A(X,L^{\phi})$, we see that (2.6) yields (2.2) for all f in $T(G)$. Now consider f in X such that (2.1) holds. By assumption, $T(G)$ is dense in X; therefore, for each $\varepsilon > o$, we can find p in $T(G)$ such that $\|f - p\|_X < \varepsilon$. We have then

$$\sup_{t,s} \|(f_t - p_t) - (f_s - p_s)\|_X < 2\varepsilon\,,$$

so that (2.6) yields:

$$\|f - p\|_{M(X,L^{\psi})^*} \leq \delta(\varepsilon)$$

with $\delta(\varepsilon) \to 0$ when $\varepsilon \to o$. This means that we can represent f as a series of trigonometric polynomials which converges absolutely in $M(X,L^{\psi})^*$. Obviously, this implies that f belongs to $A(X,L^{\phi})$, and since the norm of $M(X,L^{\psi})^*$ is equivalent on $A(X,L^{\phi})$ with the norm of $A(X,L^{\phi})$, we finally derive (2.2) from (2.6).

In the particular case $X = L^{\phi} = L^{\psi} = L^2$, then $A(L^2,L^2)$ is the algebra of absolutely convergent Fourier series on G, usually denoted $A(G)$. In that case, we obtain immediately the following "abstract" form of a theorem of S. Bernstein:

<u>Corollary 2.2</u>: Consider f in $L^2(G)$ and define

$$\forall t, s \in G \qquad d(t,s) = \|f_t - f_s\|_2 .$$

Then the condition

$$\int_0 (N(G,d;\varepsilon))^{\frac{1}{2}} \, d\varepsilon < \infty$$

implies that f is in $A(G)$ and there is an absolute constant C such that:

$$\|f\|_{A(G)} \leq C\{|\hat{f}(0)| + \int_0^D (N(G,d;\varepsilon))^{\frac{1}{2}} \, d\varepsilon\}$$

where D is defined as before.

<u>Remark 2.3</u>: Denote $\omega_2(f,t) = \|f_t - f\|_2$. Let

$$m(\varepsilon) = m(\{t \in G \mid \omega_2(f,t) < \varepsilon\}) ,$$

and $\forall u \in [0,1]$

$$\overline{\omega}_2(f,u) = \sup \{\varepsilon \mid m(\varepsilon) < u\} .$$

The function $u \to \overline{\omega}_2(f,u)$ may be considered as the "non-decreasing rearrangement" on $[0,1]$ of the function $t \to \omega_2(f,t)$. It is not hard to see that the integral

$$\int_0^D (N(G,d;\varepsilon))^{\frac{1}{2}} \, d\varepsilon$$

is equivalent to

$$\int_0^1 \frac{\overline{\omega}_2(f,u)}{u^{3/2}}\, du$$

which is analogous to the classical integral condition of S. Bernstein on the circle group. For more details concerning corollary 2.2, see [10] §4.

Remark 2.4: It is tempting (in analogy with problem 1.9) to conjecture that corollary 2.2 is best possible in the following sense:

Conjecture: Let f in $A(G)$ be such that all the functions g in $L^2(G)$ verifying

$$\forall t \in G \qquad \|g_t - g\|_2 \leq \|f_t - f\|_2$$

are also in $A(G)$. Then necessarily $\int_0 (N(G,d;\varepsilon))^{\frac{1}{2}}\, d\varepsilon < \infty$.

We could only prove (as in proposition 1.10) that $\sup_{\varepsilon > 0} \varepsilon\, (N(G,d;\varepsilon))^{\frac{1}{2}} < \infty$ is a necessary condition.

The main interest of Theorem 2.1 is that in certain cases, the exact converse is true and the metric entropy condition (2.1) characterizes the functions f in $A(L^2,L^\Phi)$. Let us first simplify the notations. For f in $L^2(G)$, we will denote simply as d the metric defined by:

$$\forall s,t \in G \qquad d(s,t) = \|f_t - f_s\|_2 = \left(\Sigma\, |\hat{f}(\gamma)|^2 |\gamma(t) - \gamma(s)|^2\right)^{\frac{1}{2}}.$$

We will write simply $N(f;\varepsilon)$ instead of $N(G,d;\varepsilon)$. Moreover, for $0 < q < \infty$, we define

$$E_q(f) = \int_0^\infty \left(\operatorname{Log} N(f,\varepsilon)\right)^{1/q} d\varepsilon .$$

Note that if $D = \inf\{\varepsilon \mid N(f,\varepsilon) = 1\}$, we have $\operatorname{Log} N(f,\varepsilon) = 0$ for $\varepsilon > D$, and also

$$D \leq \sup_{t,s} \|f_t - f_s\|_2 \leq 2\|f - \hat{f}(0)\|_2 .$$

We will denote ψ_q and ϕ_q the functions

$$\psi_q(x) = \exp|x|^q - 1 \quad \text{and} \quad \phi_q = |x|\,(1 + \mathrm{Log}(1 + |x|))^{1/q} .$$

(For $q < 2$, ψ_q is not convex; however, it is easy to see that it is equivalent to a convex function which is enough for our purposes.) As is well known, the functions (ϕ_q, ψ_q) are a couple in duality as indicated in the beginning of this section. To lighten the notations, we will write simply $A(2,\phi_q)$ instead of $A(L^2, L^{\phi_q})$ and $M(2,\psi_q)$ instead of $M(L^2, L^{\psi_q})$.

In the particular case when $\psi = \psi_q$ and $\phi = \phi_q$, Theorem 2.1 implies that there is a constant C such that: $\forall f \in L^2(G)$

$$\|f\|_{A(2,\phi_q)} \leq C \{ |\hat{f}(0)| + \int_0^D (1 + N(f,\varepsilon))^{1/q} d\varepsilon \}$$

and since $1 + N(f,\varepsilon) \leq N(f,\varepsilon)^2$ $\forall \varepsilon < D$, we have simply:

$$\|f\|_{A(2,\phi_q)} \leq 2^{1/q} C \{ |\hat{f}(0)| + E_q(f) \} , \tag{2.7}$$

for all $0 < q < \infty$.

The main result of this section shows that if $2 \leq q < \infty$, the reverse inequality also holds. The case $q = 2$ was treated in [8] (cf. also [7]):

Theorem 2.5 [8]:

There are constants A and B such that $\forall f \in A(2,\phi_2)$

$$\frac{1}{A} \{ |\hat{f}(0)| + E_2(f) \} \leq \|f\|_{A(2,\phi_2)} \leq B \{ |\hat{f}(0)| + E_2(f) \} ;$$

moreover, an element f of $L^2(G)$ belongs to $A(2,\phi_2)$ iff $E_2(f) < \infty$.

We have seen already the proof of the right hand side. For a detailed proof of the left hand side, which uses Fernique's theorem (see Theorem 1.8 above), we refer the reader to [8] or [7].

We will now explain how this result can be generalized to the case $2 < q < \infty$ as follows:

<u>Theorem 2.6:</u>

For each q with $2 < q < \infty$, there are constants A_q and B_q such that: $\forall f \in A(2, \phi_q)$

$$(2.8) \qquad \frac{1}{A_q} \{|\hat{f}(o)| + E_q(f)\} \leq \|f\|_{A(2,\phi_q)} \leq B_q \{|\hat{f}(0)| + E_q(f)\} ;$$

moreover, if f is in $L^2(G)$, then f belongs to $A(2,\phi_q)$ iff $E_q(f) < \infty$.

The right side of (2.8) has already been proved above in (2.7) without any restriction on q. The left hand side will be established below roughly by interpolating between the cases $q = 2$ and $q = \infty$. Indeed, it is natural to identify L^{ϕ_∞} with L^1 and L^{ψ_∞} with L^∞; consequently, we are led to identify $A(2,\phi_\infty)$ with L^2. On the other hand, for f in $T(G)$, we can easily check that $E_q(f) \to D$ when $q \to \infty$, and $D \leq 2\|f\|_2$. These observations show that the left hand side of (2.8) appears trivially true for $q = \infty$, while Theorem 2.5 ensures that it is true for $q = 2$.

To prove the same result for the intermediate values of q, we will need to recall some facts from the theory of Lions-Peetre interpolation. Let (A_0, A_1) be an interpolation couple of Banach spaces (i.e. two Banach spaces both continuously injected in some "larger" topological vector space; for example, L^∞ and L^1 are both injected in L^0). The norm K_t on $A_0 + A_1$ is defined as follows:

$$\forall x \in A_o + A_1, \quad \forall\, t > 0$$

$$K_t(x;A_0,A_1) = \inf\{\|u\|_0 + t\|v\|_1 \mid x = u+v\}.$$

The Banach space $[A_0,A_1]_{\theta,p}$ can then be defined (using the so-called K-method) as follows: By definition, an element x in $A_0 + A_1$ belongs to $[A_0,A_1]_{\theta,p}$ if

$$\left(\int_0^\infty (t^{-\theta} K_t(x;A_0,A_1))^p \frac{dt}{t}\right)^{1/p} < \infty$$

To endow this space with a Banach space structure, one equips it with the norm

$$\|x\|_{[A_0,A_1]_{\theta,p}} = \left(\int_0^\infty (t^{-\theta} K_t(x;A_0,A_1))^p \frac{dt}{t}\right)^{1/p}.$$

$(0 < \theta < 1,\ 1 \leq p \leq \infty)$. When applied to the interpolation couple (ℓ^1,ℓ^∞), this interpolation method yields the well known Lorentz sequence spaces $\ell^{p,q}$. The spaces $\ell^{p,q}$ can be introduced as follows: for any sequence of scalars $(\alpha_n)_{n\geq 1}$ we denote $(\alpha_n^*)_{n\geq 1}$ the non-increasing rearrangement of the sequence of their moduli $(|\alpha_n|)_{n\geq 1}$. By definition, we will say that $(\alpha_n)_{n\geq 1}$ belongs to $\ell^{p,q}$ if

$$\sum_{n=1}^{\infty} (n^{1/p}\alpha_n^*)^q \frac{1}{n} < \infty ;$$

moreover, we write

$$\|(\alpha_n)\|_{p,q} = \{\sum_{n=1}^{\infty} \alpha_n^{*q} n^{q/p-1}\}^{1/q}.$$

It is well known that $\|\ \|_{p,q}$ is a quasi-norm which is equivalent to a norm if $1 < p < \infty$ and $1 \leq q \leq \infty$. These spaces coincide with the spaces $(\ell^1,\ell^\infty)_{\theta,q}$; precisely, if $\frac{1}{p} = \frac{1-\theta}{1} + \frac{\theta}{\infty}$, then $\ell^{p,q}$ can be identified (given the equivalence of the corresponding quasi-norms) with the space $[\ell^1,\ell^\infty]_{\theta,q}$. Note that, if $p = q$, then $\ell^{p,q} = \ell^p$. By the so-called "reiteration theorem," it is also well known that, if $1 \leq q, p_0, q_0, p_1, q_1 \leq \infty$, then we have

$$[\ell^{p_0 q_0}, \ell^{p_1 q_1}]_{\theta,q} = \ell^{p_\theta q}$$

with p_θ defined by the relation $\frac{1}{p_\theta} = \frac{1-\theta}{p_0} + \frac{\theta}{p_1}$ and $0 < \theta < 1$. We will need only the following fact which we state as a lemma (for the proof of this lemma and for more details on interpolation and on Lorentz spaces, we refer the reader to [2].).

Lemma 2.7: If $2 < q < \infty$ and $\frac{1}{q} = \frac{1-\theta}{2} + \frac{\theta}{\infty}$ (i.e. $1-\theta = 2/q$), there exists a constant $C(q)$ such that $\forall\ (\alpha_n)_{n \geq 1} \in \mathbb{C}^{\mathbb{N}}$

$$\|(\alpha_n)\|_{q,1} = \sum_1^\infty \alpha_n^* n^{-1/q'} \leq C(q) \|(\alpha_n)\|_{[\ell^{2,1}, \ell^\infty]_{\theta,1}}$$

(with $\frac{1}{q} + \frac{1}{q'} = 1$).

We now proceed to rewrite the integral $E_q(f)$ as a series: Fix f in $L^2(G)$, we define

$$\sigma_n(f) = \inf \{\varepsilon > 0 \mid N(f,\varepsilon) \leq 2^n\};$$

so that on the interval $\sigma_{n+1}(f) < \varepsilon < \sigma_n(f)$, we have $2^n < N(f,\varepsilon) \leq 2^{n+1}$. Let us write for simplicity that $X \sim Y$ if we can find constants A and B (depending only on q) such that

$$\frac{1}{A} Y \leq X \leq B\, Y .$$

We then have:

$$E_q(f) \sim \sum_{n \geq 0} (n+1)^{1/q} \left(\sigma_n(f) - \sigma_{n+1}(f)\right)$$

$$= \sigma_0(f) + \sum_{n \geq 0} \sigma_{n+1}(f) \left((n+2)^{1/q} - (n+1)^{1/q}\right),$$

so that finally:

(2.9) $$E_q(f) \sim \sigma_o(f) + \sum_{k\geq 1} \sigma_k(f) k^{-1/q'} \quad \text{with} \quad \frac{1}{q} + \frac{1}{q'} = 1 .$$

We can now state the key step for the proof of the left hand side of (2.8):

<u>Theorem 2.8</u>: If $2 < q < \infty$, there is a constant $C'(q)$ such that if $\theta = 1 - \frac{2}{q}$, we have $\forall f \in [A(2,\phi_2), L^2]_{\theta,1}$

$$E_q(f) \leq C'(q) \quad \|f\|_{[A(2,\phi_2),L^2]_{\theta,1}} .$$

To prove Theorem 2.8, the following lemma (the proof of which is elementary) will be essential:

<u>Lemma 2.9</u>: Consider f in $A(2,\phi_2) + L^2$, and let $\sigma_n(f)$ be defined as above; then we have $\forall t > o$

(2.10) $$K_t\left((\sigma_{2^n}(f))_{n\geq 1} ; \ell^{2,1}, \ell^\infty\right) \leq C\, K_t\left(f; A(2,\phi_2), L^2\right)$$

where C is an absolute constant.

<u>Proof</u>: Let $f = u + v$ be a decomposition with $u \in A(2,\phi_2)$ and $v \in L^2$. Let us denote d_f the pseudo-metric defined by

$$\forall s, t \in G \quad d_f(s,t) = \|f_t - f_s\|_2 .$$

We have obviously

(2.11) $$d_f(s,t) \leq d_u(s,t) + d_v(s,t) .$$

It is easy to see that (2.11) implies

(2.12) $$N(f, 2(\varepsilon + \delta)) \leq N(u,\varepsilon)\, N(v,\delta) \quad \text{for} \quad \varepsilon,\delta > 0 .$$

Indeed, by definition of $N(u,\varepsilon)$, we can find a covering $(U_i)_{i\le N(u,\varepsilon)}$ of G by sets of d_u-diameter less than 2ε. Similarly, we can find a covering $(V_j)_{j\le N(v,\varepsilon)}$ of G by sets of d_v-diameter not exceeding 2δ. Obviously, by (2.11), the family $(U_i \cap V_j)_{\substack{i\le N(u,\varepsilon)\\ j\le N(v,\delta)}}$ is a covering of G by sets of d_f-diameter not exceeding $2(\varepsilon+\delta)$. Therefore, (2.12) follows. The inequality (2.12) implies immediately

$$\forall n \ge 1 \qquad \sigma_{2n}(f) \le 2\big(\sigma_n(u) + \sigma_n(v)\big) .$$

Clearly, one can then find numbers a_n, b_n such that $0 \le a_n \le 2\sigma_n(u)$, $0 \le b_n \le 2\sigma_n(v)$ and $\sigma_{2n}(f) = a_n + b_n$; so that we may write:

$$K_t\Big(\big(\sigma_{2n}(f)\big)_{n\ge 1} ; \ell^{2,1}, \ell^{\infty}\Big) \le \|(a_n)\|_{2,1} + t\|(b_n)\|_{\infty}$$

$$\le 2\big[\|\big(\sigma_n(u)\big)\|_{2,1} + t\|\big(\sigma_n(v)\big)\|_{\infty}\big]$$

$$\le 2 \ \Big[\sum_1^{\infty} \sigma_n(u) n^{-\frac{1}{2}} + t \sup_{n\ge 1} \sigma_n(v)\Big] .$$

By (2.9), the last expression can be majorized by

$$C' \ \{ E_2(u) + t \|v\|_2 \}$$

for some absolute constant C', and by Theorem 2.5 we conclude that

$$(2.13) \qquad K_t\Big(\big(\sigma_{2n}(f)\big)_{n\ge 1} ; \ell^{2,1}, \ell^{\infty}\Big) \le C \ \{\|u\|_{A(2,\phi_2)} + t \|v\|_2 \}$$

for some absolute constant C. Taking the infimum of the right side of (2.13) over all possible decompositions $f = u + v$, we obtain the announced result (2.10).

<u>Proof of Theorem 2.8:</u> By a suitable integration of (2.10), we obtain

$$\|\big(\sigma_{2n}(f)\big)_{n\ge 1}\|_{[\ell^{2,1},\ell^{\infty}]_{\theta,1}} \le C \|f\|_{[A(2,\phi_2),L^2]_{\theta,1}} .$$

Hence, by lemma 2.7:

$$(2.14) \qquad \sum_{n\geq 1} \sigma_{2n}(f)n^{-1/q'} \leq C(q)C \;\|f\|_{[A(2,\phi_2),L^2]_{\theta,1}}$$

Finally, we observe that

$$(2.15) \qquad \sum_{n\geq 2} \sigma_n(f)n^{-1/q'} \leq 2^{1/q} \sum_{n\geq 1} \sigma_{2n}(f)n^{-1/q'}$$

and, on the other hand, we also have

$$(2.16) \qquad \sigma_1(f) \leq \sigma_0(f) = D \leq 2\|f\|_2 \leq C''\|f\|_{[A(2,\phi_2),L^2]_{\theta,1}}$$

for some absolute constant C''. Recollecting (2.9), we conclude from (2.14), (2.15) and (2.16):

$$E_q(f) \leq C'(q) \;\|f\|_{[A(2,\phi_2),L^2]_{\theta,1}}$$

for some constant $C'(q)$ (depending only on q). q.e.d.

We now turn to the converse direction to show that $A(2,\phi_q)$ is included in $[A(2,\phi_2),L^2]_{\theta,1}$. This is much easier.

The following result plays an important rôle in the proof.

Theorem 2.10 [1]: If $2 < q < \infty$ and $\frac{1-\theta}{2} = \frac{1}{q}$, the space $L^{\phi_q}(G)$ can be identified with $[L^{\phi_2}(G), L^1(G)]_{\theta,1}$, and the corresponding norms are equivalent. (This result holds also for $o < q \leq 2$, but we only use it as stated.) For the proof, see [1]. We can then deduce by a standard argument the following fact:

Lemma 2.11: There is a constant β_q such that: $\forall f \in A(2,\phi_q)$

$$(2.17) \qquad \|f\|_{[A(2,\phi_2),L^2]_{\theta,1}} \leq \beta_q \;\|f\|_{A(2,\phi_q)} .$$

Proof: It is clearly enough to show (2.17) for $f = h * k$ with $h \in L^2(G)$ and $k \in L^{\phi_q}$. By Theorem 2.10, we know that there is a constant β_q such that

$$\|k\|_{[L^{\phi_2},L^1]_{\theta,1}} = \int_0^\infty K_t(k;L^{\phi_2},L^1)\, t^{-1-\theta}dt \le \beta_q \|k\|_{\phi_q}.$$

It is very easy to verify that $\forall t > o$

$$K_t(h * k; A(2,\phi_2), L^2) \le \|h\|_2 K_t(k;L^{\phi_2},L^1),$$

and, therefore, we obtain immediately after integration:

$$\|h*k\|_{[A(2,\phi_2),L^2]_{\theta,1}} \le \|h\|_2 \|k\|_{[L^{\phi_2},L^1]_{\theta,1}}$$

$$\le \beta_q \|h\|_2 \|k\|_{\phi_q},$$

from which (2.17) follows immediately. q.e.d.

Proof of Theorem 2.6: It remains to prove the left hand side of (2.8) Now, this follows from a combination of Theorem 2.8 and Lemma 2.11.

In the course of the proof, we have obtained:

Corollary 2.12: For $2 < q < \infty$, the space $A(2,\phi_q)$ coincides (with equivalent norms) with the space $[A(2,\phi_2),L^2]_{\theta,1}$, for $\frac{1-\theta}{2} = \frac{1}{q}$. By duality, this implies that

$$M(2,\psi_q) \approx [M(2,\psi_2),L^2]_{\theta,\infty}.$$

(Recall that, for $q = \infty$, it is natural to identify L^2 with $A(2,\infty)$ or $M(2,\infty)$.)

As immediate consequences of Theorem 2.6, we have

Corollary 2.13: For $2 \leq q < \infty$, the functional $f \to E_q(f)$ is equivalent to a norm on the subspace of $A(2,\phi_q)$ formed by the functions f such that the constant coefficient $\hat{f}(0)$ is equal to zero.

Corollary 2.14: Assume $2 \leq q < \infty$. Let f,g in $L^2(G)$ be such that:

$$\forall t,s \in G \qquad \|g_t - g_s\|_2 \leq \|f_t - f_s\|_2 .$$

Then,

$$f \in A(2,\phi_q) \Rightarrow g \in A(2,\phi_q) ;$$

moreover, there is a constant γ_q such that

$$\|g\|_{A(2,\phi_q)} \leq |\hat{g}(0)| + \gamma_q \|f\|_{A(2,\phi_q)} .$$

In the case $q = 2$, this was observed in [8]; by exactly the same arguments as in [9], we obtain:

Corollary 2.15: For $2 \leq q < \infty$, consider the space $D_q = A(2,\phi_q) \cap C(G)$ equipped with the norm

$$\|f\|_{D_q} = \|f\|_{A(2,\phi_q)} + \|f\|_{C(G)} .$$

Then: (i) D_q is a Banach algebra for the pointwise multiplication.

(ii) $A(G) \underset{\neq}{\subset} D_q \underset{\neq}{\subset} C(G)$.

(iii) All Lipschitzian functions of order 1 operate on D_q.

This means that all the algebras D_q are distinct counterexamples to the "dichotomy problem" studied in [13] and [9]. The following problem arises naturally:

Problem: Find direct proofs of the last three collaries.

Remark 2.16: It is important to emphasize that the left side of Theorem 2.8, Corollary 2.13 and Corollary 2.15 all become false if $q < 2$. Also, it is worthwhile to point that for the spaces $A(L^p, L^q)$ the result analogous to Corollary 2.12 is not true. (In that case, only one inclusion can possibly hold.)

Remark 2.17: For simplicity, we have developed the results of this section for Abelian groups only. It is quite clear, however, that all the results remain valid for a compact non Abelian group G. For details, the reader can consult Chapter 6 in [7].

It seems plausible that Corollary 2.12 admits an extension in the case of a non compact group G (for instance if $G = R$), but this does not seem entirely clear.

Note added in proof:

The recent paper [14] contains an extension of the Dudley-Fernique theorem to the p-stable case, $1 < p < 2$. This yields (with the methods of the present paper) a generalization of theorem 2.6 for the spaces $A(F\ell_p(\hat{G}), L^{\phi q}(G))$ for $\frac{1}{p} + \frac{1}{q} \leq 1$.

References

[1] C. Bennett. Intermediate spaces and the class $L\log^+ L$. Arkiv för Matematik 11 (1973) 215-228.

[2] J. Bergh, J. Löfström. Interpolation spaces. Springer Velag. Berlin, Heidelberg, New York (1976).

[3] R. M. Dudley. The size of compact subsets of Hilbert space and continuity of Gaussian processes. Journal of Funct. Analysis 1 (1967) 290-330.

[4] X. Fernique. Régularité des trajectoires des processus gaussiens. Ecole d'Ete de S^T Flour. Springer Lecture Notes n°480.

[5] A. Garsia. A remarkable inequality and the uniform convergence of Fourier series. Indiana Univ. Math Journal 25 (1976) 85-102.

[6] N. Kôno. Sample paths properties of stochastic processes. J. Math Kyoto Univ. 20 (1980) 295-313.

[7] M. B. Marcus and G. Pisier. Random Fourier series with Applications to Harmonic Analysis. Annals of Math Studies. n°101 (1981) Princeton University Press.

[8] G. Pisier. Sur l'espace des series de Fourier aléatoires presque sûrement continues. Exposé n° 17-18, Seminaire sur la géométrie des espaces de Banach. Ecole Polytechnique, Palaiseau, 1977/78.

[9] G. Pisier. A remarkable homogeneous Banach algebra. Israel J. Math 34 (1979) 38-44.

[10] G. Pisier. Conditions d'entropie assurant la continuité de certains processus et applications a l'analyse harmonique. Seminaire d'analyse fonctionnelle. Expose n° 13-14. Ecole Polytechnique, Palaiseau, 1979/80.

[11] G. Pisier. De nouvelles caractérisations des ensembles de Sidon. Advances in Maths. Supplementary Studies. (1981) Vol. 7B, p. 685.

[12] J. H. Wells, L. R. Williams. Embeddings and extensions in analysis, Springer Verlag (1975), Ergebnisse Band 84.

[13] M. Zafran. The dichotomy problem for homogeneous Banach algebras. Annals of Maths. 108 (1978) 97-105.

[14] M. B. Marcus and G. Pisier. Characterizations of almost surely continuous p-stable random Fourier series and strongly stationary processes. To appear.

SIGN-EMBEDDINGS OF L^1

H. P. Rosenthal*

Our object here is to crystallize the ideas concerning semi-embeddings of Banach spaces when the domain space is L^1. It appears that the concept of sign-embedding is more basic in this case. This notion also leads to a rather natural "soft-analysis" proof of the classical theorem of Menchoff [5] that there exists a singular probability measure on the circle with Fourier coefficients vanishing at infinity. Most of the results given here are contained in joint work of the author and J. Bourgain. They are presented in the language of semi-embeddings in [1], rather than in terms of the "crystallization" of sign-embeddings introduced here.

By L^1 we mean $L^1([0,1],\&,m)$ where $\&$ denotes the Legesgue measurable sets and m Lebesgue measure. Of course we could as well consider L^1 of any atomless separable finite measure space. Through out, "operator" means "bounded linear map." Let X and Y be Banach spaces and $T: X \to Y$ a given operator. We say that T is a semi-embedding if T is one-one and $T(Ba(X))$ is closed, where $Ba(X) = \{x \in X: ||x|| \leq 1\}$. (This concept was first introduced in [4].) T is an embedding if there is a $\delta > 0$ so that $||Tx|| \geq \delta||x||$ for all $x \in X$. If $X = L^1$, we say that T is a sign-embedding if T is one-one and there is a $\delta > 0$ so that $||Tx|| \geq \delta||x||$ whenever x is a 1,0,-1-valued function (which we think of as a "sign"). We shall present variations of the arguments in [1] which yield that if L^1 semi-embeds in Y, L^1 sign-embeds in Y. The main unsolved problem in this regard is as follows: If L^1 sign-embeds in Y, does L^1 embed in Y? By the results of Enflo-Starbird [3], the answer is affirmative if Y itself embeds in L^1. It is also affirmative if Y

*This research was supported in part by NSF-MCS-8002393 at the University of Texas at Austin.

is isomorphic to a dual space. The local problem has been solved in the affirmative by John Elton [2]. Further remarks concerning our general knowledge about sign-embeddings of L^1 will be made below.

We begin with a result presented in terms of Banach-valued martingales. The result allows us to see the connection between sign-embeddings and semi-embeddings of L^1, and also the basic result of [1] concerning the connection between semi-embeddings and the Radon-Nikodym property (the RNP).

Lemma 1. Let X and Y be Banach spaces, $T: X \to Y$ a given operator and K a closed bounded separable convex subset of X. Suppose that $T(K)$ is closed and $T|K$ is one-one. Then if (f_n) is a K-valued martingale so that (Tf_n) converges almost everywhere, (f_n) converges almost everywhere.

(This is given as Lemma 2.10 of [1]. For the sake of completeness, we give the proof here.)

Proof. For $\mathcal{a}$ a σ-subalgebra of $\&$, let $E_{\mathcal{a}}$ denote conditional expectation with respect to $\mathcal{a}$. We may choose an increasing sequence $(\mathcal{a}_n)$ of σ-subalgebras of $\&$ so that for all n, f_n is $\mathcal{a}_n$ measurable with $E_{n-1}f_n = f_{n-1}$ for $n > 1$, where $E_n = E_{\mathcal{a}_n}$ for all n. We may and shall assume that $\&$ equals the smallest complete σ-algebra containing all the $\mathcal{a}_n$'s. By assumption there is a Bochner integrable g so that $Tf_n \to g$ a.e. (Of course (Tf_n) is also a martingale with respect to $(\mathcal{a}_n)$.) Since TK is closed, we may assume that g is valued in TK everywhere. Now define $f = T^{-1}g$. K is a Polish space and $T|K$ is one-one continuous. By a classical theorem of Lusin, if U is a Borel subset of K, TU is a Borel set, hence $f^{-1}(U) = g^{-1}(TU)$ is Lebesgue measurable. Thus f is measurable and so Bochner integrable. By the Doob martingale convergence theorem, it follows that $E_n f \to f$ a.e. We need only show that $E_n f = f_n$ a.e. for all n. In turn, since T is one-one, it suffices to show that $TE_n f = Tf_n$ a.e. for all n. But fixing n, $TE_n f = E_n Tf = \lim_{m\to\infty} E_n Tf_m = \lim_{m\to\infty} TE_n f_m = Tf_n$.

Corollary 2. Let X, Y, T and K be as in Lemma 1. Then K has the RNP if and only if TK has the RNP.

Proof. It is a known fact that K has the RNP if and only if every K-valued martingale converges almost-everywhere. Corollary 2 thus follows immediately from Lemma 1 and this fact.

Of course Corollary 2 yields the basic structural result discussed in greater detail in [1]; if X is separable and X semi-embeds in a space with the RNP, then X has the RNP.

We pass now to the connection between sign-embeddings and semi-embeddings of L^1. We first require the following equivalence, allowing a formal weakening of the hypotheses in the definition of sign-embedding.

Lemma 3. Let X be a Banach space so that there exists an operator $T: L^1 \to X$ with the following property: there is a $\delta > 0$ so that $\|Tf\| \geq \delta$ for all $f \in L^1$ with $\int f\,dm = 0$ and $|f| \equiv 1$. Then L^1 sign-embeds in X.

Proof. We first observe that there exists a measurable set E of positive measure so that $\|Tf\| \geq \delta\, \|f\|$ for all f vanishing outside E with f 1,0,-1-valued and $\int f\,dm = 0$. Indeed, were this false, then a measure exhaustion argument yields a sequence $f_1, f_2, \ldots$ of disjointly supported 1,0,-1-valued functions of mean zero with $\Sigma |f_i| = 1$ a.e. and $\|Tf_i\| < \delta\, \|f_i\|$ for all i. Then setting

$f = \sum_{i=1}^{\infty} f_i$, we have that $|f| \equiv 1$ a.e., $\int f dm = 0$, yet $\|Tf\| \leq \Sigma \|Tf_i\| < \delta\Sigma\|f_i\| = \delta$. Next, since $[-1,1]$ and E are measure isomorphic (with E endowed with $\mathcal{B} \cap E$ under the measure $2m/m(E)$), we may choose an operator $S: L^1[-1,1] \to L^1(E)$ so that for all $f \in L^1[-1,1]$, $2\|Sf\| = m(E)\|f\|$, $2\int_E Sfdm = m(E)\int_{[-1,1]} fdm$ and Sf is $1,0,-1$-valued if f is. Finally, let $O: L^1[0,1) \to L^1[-1,1)$ be defined by

$$(Of)(x) = f(x) \quad \text{if} \quad x \geq 0\ ,$$

$$(Of)(x) = -f(-x) \quad \text{if} \quad x < 0\ .$$

Now setting $\delta' = \delta\, m(E)$, then the map $U = TSO$ has the property that $\|Uf\| \geq \delta'\|f\|$ for all $1,0,-1$-valued f on $[0,1]$.

The mapping U is "almost" the desired sign-embedding; however U may not be one-one (unless of course T is one-one). A standard argument now yields the existence of an atomless σ-subalgebra $\mathcal{a}$ of $\mathcal{B}$ such that $U\, L^1(m|\mathcal{a})$ is one-one. Since there is an isometry $V: L^1 \to L^1(m|\mathcal{a})$ with Vf $1,0,-1$-valued whenever f is, UV is the desired sign-embedding of L^1 into X.

The standard argument is as follows: Let $E_1 = [0,1]$, $h_0 \equiv 1$, and choose f in the unit ball of X^* with $f(Uh_0) = \|Uh_0\|$. Set $F_0 = \{f\}$. Now construct by induction on n, measurable sets E_n and finite subsets F_n of the unit ball of X^* so that for all $n = 1,2,\ldots$

(a) $E_n = E_{2n} \cup E_{2n+1}$ and $E_{2n} \cap E_{2n+1} = \emptyset$.

(b) $m(E_{2n}) = \frac{1}{2} m(E_n)$.

(c) Setting $h_n = \chi_{E_{2n}} - \chi_{E_{2n+1}}$ and $h_0 = 1$ and letting Y_n equal the linear span or the h_i's for $0 \leq i \leq n$, then $\sup_{f \in F_n} |f(x)| \geq \frac{1}{2}\|x\|$ if $x \in TY_n$.

(d) $f(Uh_n) = 0$ for all $f \in F_{n-1}$ and $F_{n-1} \subset F_n$. We then let $\mathcal{a}$ equal the σ-algebra generated by the sets $\{E_j: j = 1,2,\dots\}$. It follows that $(h_j)_{j=0}^{\infty}$ is a basis for $L^1(\mathcal{a})$ and $(Uh_j)_{j=0}^{\infty}$ is a basic sequence in X so $U|L^1(\mathcal{a})$ is one-one.

To construct the h_j's by induction, fix $n > 0$ and suppose h_i has been constructed for $0 \le i < n$ and also F_{n-1} has been constructed satisfying (c). Now it follows that the set E_n has already been constructed. By the Liapunoff convexity theorem, there exists a measurable set $E = E_{2n}$ so that $E \subset E_n$, $m(E) = \frac{1}{2} m(E_n)$ and $\int_E f dm = \frac{1}{2} \int_{E_n} f dm$ for all $f \in F_{n-1}$. We now simply set $E_{2n+1} = E \sim E_{2n}$. (Of course (h_n) has the same distribution as the standard Haar basis for L^1, normalized in L^{∞}.) Finally, choose F_n satisfying (c) with $F_{n-1} \subset F_n$, F_n a finite subset of $Ba(X^*)$. This completes the proof of Lemma 3.

We are now prepared to relate sign-embeddings and semi-embeddings of L^1. We let $\mathcal{p}$ denote the set of all (equivalence classes of) probability densities in L^1. That is, $f \in \mathcal{p}$ if and only if $f \ge 0$ a.e. and $\int f dm = 1$.

<u>Theorem 4</u>. Let X be a Banach space with the following property: there exists a one-one operator $T: L^1 \to X$ and a closed bounded convex set W with TW closed and $\mathcal{p} \subset W$. Then L^1 sign-embeds in X.

<u>Proof</u>. Let T and W as in the statement of the theorem. By Lemma 3 it suffices to prove that there exists a set E of positive measure and a $\delta > 0$ so that $\|Tf\| \ge \delta$ for all measurable f with $\int f dm = 0$ and $|f| = \chi_E$. (χ_E is the characteristic function of E; $\chi_E(x) = 1$ if $x \in E$, $\chi_E(x) = 0$ if $x \notin E$.) Suppose there is no

such set E. Set $h_0=1$ and $E_1 = [0,1]$. We again define a sequence of measurable sets (E_n) satisfying (a), (b) of the previous lemma and functions (h_n) with $h_n = \chi_{E_{2n}} - \chi_{E_{2n+1}}$ and $\|Th_n\| < \|\frac{h_n}{2^n}\|$. for all n. Indeed, suppose h_i defined for $0 \leq i < n$. Thus E_n is defined; choose h_n so that $\int h_n dm = 0$, $|h_n| = \chi_{E_n}$ and $\|Th_n\| < \frac{\|h_n\|}{2^n}$. Now set $E_{2n} = \{\omega: h_n(\omega) = 1\}$ and $E_{2n+1} = \{\omega: h_n(\omega) = -1\}$.

We now define a martingale valued in $\mathcal{D}$ by $\vec{f}_n(\omega) = \sum_{i=0}^{n-1} \frac{h_i}{\|h_i\|} h_i(\omega)$, $n = 1,2,\ldots$. Now for any $\omega \in [0,1]$ there are infinitely many n so that $|h_n(\omega)| = 1$; thus $\|\vec{f}_{n+1} - \vec{f}_n(\omega)\| = 1$ for infinitely many n, so $(\vec{f}_n)$ is nowhere convergent. However $(T\vec{f}_{n+1} - T\vec{f}_n)(\omega) = \frac{Th_n}{\|h_n\|} \cdot h_n(\omega)$; thus $\|(T\vec{f}_{n+1} - T\vec{f}_n)(\omega)\| < \frac{1}{2^n}$ for all n, so $(T\vec{f}_n)(\omega)$ converges for all ω. This contradicts Lemma 1, completing the proof.

Of course, Theorem 4 yields immediately that if L^1 semi-embeds in X then L^1 sign-embeds in X. In fact, it is proved in [1] that if L^1 G_δ-embeds in X, then L^1 sign-embeds in X. ($T: Y \to X$ is called a G_δ-embedding of TK is a G_δ for all closed bounded sets K. It is observed in [1] that if Y is separable and T is a semi-embedding, T is a G_δ-embedding.) In fact the proof of Theorem 4 can be sharpened to yield that if W is a closed bounded convex set with $\mathcal{D} \subset W$ and if $T: L^1 \to X$ is a given one-one operator with TW a G_δ, then L^1 sign-embeds in X. We formulated the desired construction as follows:

Lemma 5. Let X be a Banach space such that L^1 does not sign-embed in X, let $T: L^1 \to X$ be a given operator, and let G be a G_δ-subset of X with $T\rho$ G. Then there exists a martingale (f_n) valued in ρ such that (f_n) converges nowhere, yet (Tf_n) converges everywhere to a function g valued in G.

The result stated immediately above Lemma 5 now follows from Lemma 5 and the proof of Lemma 1. Indeed, suppose L^1 does not semi-embed in X yet $T: L^1 \to X$ is one-one with TW a G_δ. Choose (f_n) as in Lemma 5 with $Tf_n \to g$ a.e., g valued in TW. But now the proof of Lemma 1 shows that $f_n \to T^{-1}g$ a.e., a contradiction. (The crucial point is that $T^{-1}g$ is indeed defined.)

Proof of Lemma 5. Choose open sets $U_1 \supset U_2 \supset \ldots$ with $G = \bigcap_{n=1}^{\infty} U_n$. Set $h_0 = 1$ and $E_1 = [0,1]$ as before. Choose B_1 an open ball in X centered at $T1$ with radius at most $\frac{1}{2}$ with $B_1 \subset U_1$. We now choose by induction measurable sets E_n and open balls B_n in X so that the following hold for all n:

(a) $E_n = E_{2n} \cup E_{2n+1}$ and $E_{2n} \cap E_{2n+1} = \emptyset$

(b) $m(E_{2n}) = \frac{1}{2} m(E_n)$

(c) B_n has radius at most $\frac{1}{2^n}$ and B_n is centered at $T(\chi_{E_n}/m(E_n))$

(d) $B_n \subset U_n$ and $\bar{B}_{2n+1} \cup \bar{B}_{2n} \subset B_n$.

The sets E_n and balls B_n are constructed in pairs. Suppose k is a positive integer and the sets E_i and balls B_i have been constructed for all $i < 2k$. Let $a = m(E_k)$, set $x = T(\chi_{E_k}/a)$ and let r be the radius of B_k. Thus $B_k = \{y \in X: \|y-x\| < r\}$. Now set $\tau = \min\{\frac{r}{2}, 2^{-(2k+1)}\}$ and choose a function h with $|h| = \chi_{E_k}$,

$\int h\,dm = 0$ and $\|T(h/a)\| < \tau$. Set $E_{2k} = \{x: h(x) = 1\}$, and $E_{2k+1} = \{x: h(x) = -1\}$. Let $y = T(h/a)$. Now let B_{2k} be an open ball of radius at most τ centered at $x+y$ so that $B_{2k} \subset U_{2k}$; also let B_{2k+1} be an open ball of radius at most τ centered at $x-y$ so that $B_{2k+1} \subset U_{2k+1}$. We note that $T(\chi_{E_{2k}}/m(E_{2k})) = x+y$ and $T(\chi_{E_{2k+1}}/m(E_{2k+1})) = x-y$; hence $\bar{B}_{2k} \cup \bar{B}_{2k+1} \subset B_k$.

This completes the construction of the E_n's and B_n's. As before, set $h_n = \chi_{E_{2n}} - \chi_{E_{2n+1}}$ and $\vec{f}_n(\omega) = \sum_{i=0}^{n-1} \frac{h_i}{\|h_i\|} h_i(\omega)$ for all n and ω. Of course $(\vec{f}_n)$ is valued in $\mathcal{D}$ and converges nowhere. Now suppose $\omega \in [0,1]$. There exists a unique sequence of integers $1 = n_1 < n_2 < n_3 < \cdots$ so that for all i, $\omega \in E_{n_i}$ and $m(E_{n_i}) = \frac{1}{2^{i-1}}$. Now it follows that for all i, $n_{i+1} = 2n_i$ or $n_{i-1} = 2n_i + 1$, hence $\bar{B}_{n_{i+1}} \subset B_{n_i}$ and since the radii of the B_j's go to zero, there is a unique point $z \in \bigcap_{i=1}^{\infty} B_{n_i} \subset G$ by (d). Fixing i and j with $n_i < 2j \le n_{i+1}$, then $\vec{f}_j(\omega) = \chi_{E_{n_i}}/m(E_{n_i})$ hence $T\vec{f}_j(\omega) \in B_{n_i}$. It follows that $T\vec{f}_j(\omega) \to z$ as $j \to \infty$; hence $(T\vec{f}_j)$ converges everywhere to a function valued in G, completing the proof.

We proceed now to the discussion of the fact that L^1 does not sign-embed in c_0 along with some consequences of this fact in harmonic analysis. It should be pointed out that considerably deeper results are known. For example, in [1] it is proved that if $T: L^1 \to X$ is a sign-embedding then there is a subspace Y of L^1 with y isomorphic to ℓ^1 so that $T|Y$ is an isomorphism. Hence ℓ^1 embeds in X, so of course X could not be isomorphic to c_0. Also, the author shows in [6] that L^1 does not sign-embed in any space with an unconditional basis.

<u>Proposition 6</u>. L^1 does not sign-embed in c_o.

(This is given as Lemma 2.8 of [1]; for the sake of completeness, we sketch the argument here.) Let $T: L^1 \to c_o$ be a given operator and $\varepsilon > 0$. We construct a function r with $|r| = 1$ a.e. and $\int r dm = 0$ so that $\|Tr\| < \varepsilon$. Assume without loss of generality that $\|T\| \leq 1$. Choose k with $\frac{1}{k} < \frac{\varepsilon}{2}$ and set $I_j = [\frac{j-1}{k}, \frac{j}{k})$ for all j with $1 \leq j \leq k$. We shall choose r of the form $r = \sum_{j=1}^{k} r_j$ where for all j

(*) $$|r_j| = \chi_{I_j} \quad \text{and} \quad \int r_j dm = 0 .$$

Choose r_1 arbitrarily satisfying (*) then choose N_1 so that $|Tr_1(n)| < \frac{\varepsilon}{4}$ for all $n \geq N_1$. Now using the Liapunoff convexity theorem, choose by induction integers $N_2, \ldots, N_k$ and functions $r_2, \ldots, r_k$ so that for all j, $2 \leq j \leq k$,

(a) $N_{j-1} < N_j$ and r_j satisfies (*)

(b) $Tr_j(n) = 0$ if $n < N_{j-1}$ and

(c) $|Tr_j(n)| < \frac{\varepsilon}{2^{j+1}}$ if $n \geq N_j$.

This accomplished, it follows that the Tr_j's are "almost disjointly supported" in c_o; in fact $\|Tr\| = \|\Sigma Tr_j\| \leq 2 \max_j \|Tr_j\| < \varepsilon$.

We treat finally the promised application to harmonic analysis. We first note that if X and Y denote the complex Banach spaces L^1 and c_o respectively, then if $T: X \to Y$ is a given one-one operator; $T\mathcal{P}$ is not closed. Indeed, Y is still isomorphic to the real space c_o when Y is regarded as a Banach space over the reals, so this follows from Proposition 6 and Theorem 4.

We introduce the following notation: for G a compact metrizable abelian group with dual group Γ and μ a finite complex Borel measure on G, the Fourier transform $\hat{\mu}$ of μ is defined by $\hat{\mu}(\gamma) = \int_G \gamma(x)\,d\mu(x)$ for all $\gamma \in \Gamma$. $M(G)$ denotes the set of all complex Borel measures on G; the weak*-topology on $M(G)$ refers to the $C(G)$-topology. For $\mu \geq 0$ in $M(G)$ and $f \in L^1(\mu)$, define $f\cdot\mu$ by $(f\cdot\mu)(E) = \int_E f\,d\mu$ for all Borel sets E.

Corollary 7. Let G be a compact metrizable abelian group with dual group Γ and ν a Borel probability measure on G with $\hat{\nu} \in c_o(\Gamma)$. Then there is a Borel probability measure λ with $\hat{\lambda} \in c_o(\Gamma)$ so that $\lambda \perp \nu$ and λ is in the weak*-closure of a bounded subset of $L^1(\nu)$.

Applying this to the circle group and $\nu = m$ = Lebesgue measure, we obtain the classical Menchoff theorem that there exists a Borel probability measure μ on $[0,2\pi)$ with $\hat{\mu} \in c_o(\mathbb{Z})$ and $\hat{\mu} \perp m$. If λ is absolutely continuous with respect to μ, then also $\hat{\lambda} \in c_o(\mathbb{Z})$. Hence by the regularity of μ, there is also a probability measure λ supported on a closed set K of measure zero with $\hat{\lambda} \in c_o(\mathbb{Z})$; this is the form in which Menchoff originally stated his result [5].

Proof of Corollary 7. As we remarked above, if λ is absolutely continuous with respect to ν, λ a complex Borel measure, then also $\hat{\lambda} \in c_o(\Gamma)$. Indeed there exist trigonometric polynomials P_n so that $P_n\cdot\nu \to \lambda$ in measure norm, whence $(P_n\cdot\nu)^\wedge \to \hat{\lambda}$ uniformly. Now define $T: L^1(\nu) \to c_o(\Gamma)$ by $Tf = (f\cdot\nu)^\wedge$ for all $f \in L^1(\nu)$. Then T is a one-one operator, hence $T(p(\nu))$ is not closed by Proposition 6 and Theorem 4, as noted above. ($p(\nu)$ denotes the set of probability densities in $L^1(\nu)$.) Thus we may choose a sequence (f_n) in $p(\nu)$ so that (Tf_n) converges in $c_o(\Gamma)$ to a function g with $g \notin Tp(\nu)$. But then $(f_n\cdot\nu)$ converges weak* to a Borel probability measure μ; it follows that $\hat{\mu} = g$. Now μ cannot be absolutely continuous with respect to ν, or else $\mu = f\cdot\nu$ for some $f \in p$. Hence there exists a positive measure β with $\beta \neq 0$ and a positive $f \in L^1(\nu)$ so that $\beta \perp \nu$ and $\mu = f\cdot\nu+\beta$. Then since $(f\cdot\nu)^\wedge \in c_o(\Gamma)$, $\hat{\beta} \in c_o(\Gamma)$ also. Of course $\mu - f\cdot\nu$ is in the weak*-closure of a bounded subset of $L^1(\nu)$, so $\lambda = \beta/\|\beta\|$ is also, completing the proof.

References

1. J. Bourgain and H. P. Rosenthal, "Applications of the theory of semi-embeddings to Banach space theory," submitted to the Journal of Functional Analysis.

2. John Elton, "Sign-embeddings of ℓ_1^n," to appear.

3. P. Enflo and T. W. Starbird, "Subspaces of L^1 containing L^1," Studia Math. 65(1979), 203-225.

4. H. P. Lotz, N. T. Peck and H. Porta, "Semi-embeddings of Banach spaces," Proc. Edinburgh Math. Soc. 22(1979), 233-240.

5. D. Menchoff, "Sur l'unicite du developpment trigonometrique," Comptes Rendus de l'Academie des Sciences 163(1916), 433-436.

6. H. P. Rosenthal, "A new stopping-time Banach space," in preparation.

EXPOSED POINTS FOR DUALS OF SEPARABLE FRÉCHET SPACES

S. J. Sidney*

We offer a simple proof of the following.

<u>Theorem</u>. Let E be a separable Fréchet space and K a non-empty convex weak* compact subset of its topological dual E^*. Then K is the weak* closed convex hull of its set of weak* exposed points.

For Banach spaces the theorem is certainly known, and is implicit in the deep paper of Asplund [1], in which a proof, readily adaptable to the present situation, is given of the following result attributed to E. Bishop: If E is a Banach space with (norm) separable dual and K is a non-empty convex weak* compact subset of E^*, then K is the weak* closed convex hull of its set of weak* strongly exposed points. It is not yet clear whether the Bishop-Asplund result holds for Fréchet spaces. Indeed, the result cannot even be properly formulated without first settling on a suitable choice of topology on E^* stronger than the weak* topology; one possible choice is the topology of uniform convergence on bounded subsets of E.

If E is a topological vector space and K a subset of E^*, then $x \in E$ is said to expose f in K if $f \in K$ and $\operatorname{Re} g(x) < \operatorname{Re} f(x)$ for all $g \in K$ other than f; those f which are so exposed by elements of E are the weak* exposed points of K, and form a (generally proper) subset of the set of extreme points of K. Though we will not need it, we note that if E is Banach, the notion of (weak*) strong exposure is obtained by adding to (weak*) exposure the requirement that $g_n \in K$, $\operatorname{Re} g_n(x) \to \operatorname{Re} f(x)$ as $n \to \infty$ implies $\|g_n - f\| \to 0$ as $n \to \infty$.

---

*Department of Mathematics, University of Connecticut, Storrs, CT 06268.

If E is a euclidean space, one obtains exposed points by intersecting K with appropriate spheres. The procedure of our proof will be to embed E^* in Hilbert space, the embedding being weak* continuous on K, and then to intersect (the image of) K with various spheres. We now present the details.

Proof of the Theorem. As usual, it suffices to prove that if L is any non-empty convex weak* compact proper subset of K, then $K \setminus L$ contains a weak* exposed point of K. To do this, choose a sequence $(x_n)_1^\infty$ of points of E whose linear span is dense in E, and a sequence $(k_n)_1^\infty$ of positive constants such that if $(t_n)_1^\infty$ is any sequence of scalars satisfying $|t_n| \le k_n$, the sum $\sum_1^\infty t_n x_n$ converges in E. [Remark: The proof, and so the theorem, remains valid for any topological vector space E for which such a choice of vectors x_n and positive numbers k_n is possible.] For every $f \in E^*$ the series $\Sigma\, t_n f(x_n)$ converges whenever $|t_n| \le k_n$, so $\Sigma\, k_n f(x_n)$ is absolutely convergent. Thus there is a linear injection $E^* \to H = \ell^2$ given by $f \to Tf = \big(k_n f(x_n)\big)_1^\infty$. If we arrange (by shrinking the k_n if necessary) that $\Sigma\, k_n^2 \sup\{|f(x_n)|^2 : f \in K\} < \infty$, then the restriction of T to K is weak*-to-norm continuous. For simplicity we shall identify $f \in E^*$ with $Tf \in H$ and shall denote by $N(f)$ and $\langle f,g \rangle$ the norm and inner product in H.

Choose $h_o \in K \setminus L$ and then choose $g_o \in L$ so that $d \equiv N(g_o - h_o) \le N(g - h_o)$ for all $g \in L$. Set $M = \max\{N(g - g_o) : g \in L\} < \infty$, choose $C > 0$ large enough so that $(2C + 1)d^2 - M^2 > 0$, set $h_1 = g_o + C(g_o - h_o)$, and choose $f_o \in K$ so that $N(f_o - h_1) \ge N(f - h_1)$ for all $f \in K$. We shall prove that $f_o \notin L$ and that f_o is weak* exposed in K.

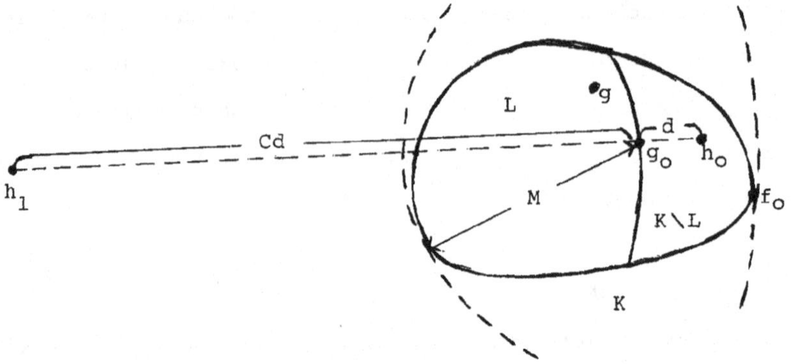

Let $g \in L$. If $0 \le t \le 1$, $g_t = (1-t)g_o + tg \in L$ so $d^2 \le N^2(g_t - h_o) =$ $= t^2 N^2(g - g_o) + 2t\,\mathrm{Re} \langle g - g_o, g_o - h_o\rangle + d^2$, hence $\mathrm{Re} \langle g - g_o, g_o - h_o\rangle \ge 0$. Thus $N^2(h_o - h_1) - N^2(g - h_1) = (C+1)^2 d^2 - N^2\big((g - g_o) + (g_o - h_1)\big) =$

$= (C+1)^2 d^2 - N^2(g - g_o) - 2\mathrm{Re} \langle g - g_o, g_o - h_1\rangle - N^2(g_o - h_1) \ge$

$\ge (C+1)^2 d^2 - M^2 + 2C\mathrm{Re} \langle g - g_o, g_o - h_o\rangle - C^2 d^2 \ge (2C+1)d^2 - M^2 > 0$, so $N(g - h_1) < N(h_o - h_1) \le N(f_o - h_1)$ and $g \ne f_o$. Therefore $f_o \notin L$.

Let $x = \sum k_n^2 \overline{(f_o - h_1)(x_n)}\, x_n \in E$; this makes sense since $k_n \overline{(f_o - h_1)(x_n)} \to 0$ as $n \to \infty$. If $f \in K$ then $|(f - h_1)(x)| = |\langle f - h_1, f_o - h_1\rangle| \le N(f - h_1)N(f_o - h_1) \le N^2(f_o - h_1) =$ $= (f_o - h_1)(x)$ with equality iff $f - h_1 = s(f_o - h_1)$ with $|s| = 1$. It follows readily that x exposes $f_o - h_1$ in $K - h_1$, so exposes f_o in K.

Reference

[1] E. Asplund, Fréchet differentiability of convex functions, Acta Math. 121 (1968) 31-47.

A STRONG CONVERGENCE THEOREM FOR $H^1(\mathbb{T})$

Brent Smith

$L^1(\mathbb{T})$ denotes the Lebesgue integrable functions on the circle group, $\mathbb{T}$. $H^1(\mathbb{T}) = \{f \in L^1(\mathbb{T}) : \hat{f}(-n) = 0 \ \forall n \in \mathbb{N}\}$. $S_n f = D_n * f$ where D_n is the Dirichlet kernel. It is easy to construct $f \in L^1(\mathbb{T})$ for which $\lim_{n\to\infty} \|S_n f\|_1 = +\infty$; all that is required is to build f so that $|\hat{f}(n)| > \frac{\theta(n)}{\log n}$ where $\theta(n) \to \infty$. However, the situation for $H^1(\mathbb{T})$ is different.

Theorem. For $f \in H^1(\mathbb{T})$.

$$\frac{1}{\log p} \sum_{k=1}^{p} \frac{\|S_k(f)\|_1}{k} \to \|f\|_1 \quad \text{as} \quad p \to \infty .$$

Proof. For $f \in L^1(\mathbb{T})$. Let

$$d_n(f) = \frac{\|S_{2^n+1}(f)\|_1 + \|S_{2^n+2}(f)\|_1 + \dots + \|S_{2^{n+1}}(f)\|_1}{2^n} .$$

We want to show if $f \in H^1(\mathbb{T})$, then

$$\frac{d_1(f) + \dots + d_N(f)}{N} \to \|f\|_1 \quad \text{as} \quad N \to \infty .$$

It suffices to show this for the real and imaginary parts of f. Thus we look at $\frac{f+\bar{f}}{2}$ and rename this real object f. As a consequence of H^1 - BMO duality, [G], f has an atomic decomposition $f = \Sigma \alpha_i a_i$ where $\Sigma|\alpha_i| < \infty$ and each a_i is an atom. a_i is an atom means supp a_i sn interval I_i, $\sup|a_i| < \frac{1}{|I_i|}$, and $\hat{a}_i(0) = 0$. Let $S_n'(f) = S_n(f) - f$.

Let

$$d_n'(f) = \frac{\|S'_{2^n+1}(f)\|_1 + \cdots + \| S'_{2^{n+1}}(f)\|_1}{2^n} .$$

We accomplish our goal by showing $\frac{d_1'(f) + \cdots + d_n'(f)}{n} \to 0$. We do this by showing $E_m'(f) = \frac{d_{m+1}'(f) + \cdots + d_{2m}'(f)}{m} \to 0$ as $m \to \infty$.

Our main tool is contained in this

<u>Lemma</u> (Charles Fefferman). For a = an I atom (that is, a is supported in I and $\sup|a| < \frac{1}{|I|}$) we have

$$\|S'_n(a)\|_1 \leq |\hat{a}(n)| \, |\log|I|| + o(n)$$

where $o(n) \leq$ a universal constant.

<u>Lemma Proof</u>. Since

$$(D_n * f)(x) = \int_{-\pi}^{\pi} \frac{\sin(n+\frac{1}{2})(x-y)}{\sin(\frac{x-y}{2})} f(y) \frac{dy}{2\pi} \quad \text{and}$$

$$(D_n^* * f)(x) = \int_{x-\pi}^{x+\pi} \frac{\sin n(x-y)}{(x-y)} f(y) \frac{dy}{\pi} \quad \text{differ by}$$

a $o(1)$ (as $n \to \infty$) term that is uniform in x we interchange these expressions at will. Center I at the origin. We estimate

$$\int_{2I} |D_n * a - a| \frac{dx}{2\pi} + \int_{[-\pi,\pi]\setminus 2I} |D_n^* * a - a| \frac{dx}{2\pi} = A_1 + A_2 .$$

We estimate A_1. Here we use the facts $\|a\|_2 \leq \sqrt{\frac{I}{2\pi}\frac{1}{|I|}}$ and $\|D_n * a - a\|_2 \to 0$ with n. So $\int_{2I} |D_n * a - a| \frac{dx}{2\pi} \leq \sqrt{\frac{I}{\pi}} \cdot \|D_n * a - a\|_2$.

We estimate A_2.

$$\int_{[-\pi,\pi]\setminus 2I} \left| \int \frac{\sin n(x-y)}{(x-y)} \frac{dy}{\pi} \right| \frac{dx}{2\pi} =$$

$$= \int_{[-\pi,\pi]\setminus 2I} \left| \int \frac{\sin n(x-y)}{x} a(y)\left(1 + \frac{y}{x} + \left(\frac{y}{x}\right)^2 + \cdots\right) \right| \frac{dy}{\pi} \frac{dx}{2\pi} .$$

where we are utilizing supp $a \subset I$ and so $\left|\frac{y}{x}\right| \leq \frac{1}{2}$ for $y \in$ supp a.

$$\leq \frac{2}{\pi} \hat{a}(n) \,||\log|I||| + \int_{[-\pi,\pi]\setminus 2I} \left| \int \left(\frac{y}{x^2} + \frac{y^2}{x^3} + \cdots\right) \sin n(x-y) a(y) \frac{dy}{\pi} \right| \frac{dx}{2\pi} .$$

The 2nd integral is under control since it is $\leq$.

$$\sum_{k\geq 1} \int_{[-\pi,\pi]\setminus 2I} \frac{1}{|x|^{k+1}} \left| \int_I y^k a(y) \sin n(x-y) \frac{dy}{\pi} \right| \frac{dx}{2\pi} \quad \text{and}$$

$|(y^k a(y))^\wedge(n)| = o(n).\ |I|^k$ where $o(n) \leq$ universal constant.

To complete the Theorem proof we will use the observations

(i) $\Sigma|\hat{a}(n)|^2 \leq \frac{1}{2\pi|I|}$; (ii) $|\hat{a}(n)| \leq \frac{n|I|}{2\pi}$ where a is an I atom.

(ii) is a consequence of $\hat{a}(0) = 0$.

We group the atoms dyadically.

$$f = \sum_{i,j} \alpha_{j,i}\, a_{j,i} ,$$

where $a_{j,i}$ is an I atom with $2^{-i-1} \leq |I| < 2^{-i}$. Let

$$A_i = \sum_j \alpha_{j,i} a_{j,i} ; \quad \beta_i = \sum_j |\alpha_{j,i}| .$$

We show:

$$E'_m(f) \leq 14\ (\beta_{m+1} + \cdots + \beta_{2m}) + o(m) .$$

In our calculations we will ignore the $o(n)$ term of the lemma.

The argument splits into 3 parts. We consider the past $A_1, \ldots, A_m$ contribution, the future $A_{2m+1}, A_{2m+2}, \ldots$ contribution and the present $A_{m+1}, \ldots, A_{2m}$ contribution to $E'_m(f)$. The past and future contribute only a $o(m)$ term.

The past starts at A_m. We estimate the size of $d'_{m+1}(A_m), \ldots, d'_{2m}(A_m)$. $d'_{m+1}(A_m)$ is largest when we utilize the maximal disparity between $\sum_{2^{m+1}+1}^{2^{m+2}} |\hat{A}_m(k)|$ and $\sum_{2^{m+1}+1}^{2^{m+2}} |\hat{A}_m(k)|^2$. By the lemma, $\|S'_k(A_m)\|_1 \leq |\hat{A}_m(k)|(m+1)$. So by (i) $d'_{m+1}(A_m) \leq \beta_m(m+1)$. Continuing while keeping in mind the continual doubling we have $d'_{m+1}(A_m) + \cdots + d'_{2m}(A_m) \leq (m+1)(\beta_m + \frac{1}{\sqrt{2}}\beta_m + \cdots + (\frac{1}{\sqrt{2}})^{m-1}\beta_m)$. Altogether $E'_m(A_m) \leq \frac{m+1}{m}\frac{1}{\sqrt{2}-1}\beta_m$. Now summing over the past

$$E'_m(A_1 + \cdots + A_m) \leq \frac{1}{m}\frac{1}{\sqrt{2}-1}\left((m+1)\beta_m + m\frac{1}{\sqrt{2}}\beta_{m-1} + \cdots + 2(\frac{1}{\sqrt{2}})^{m-1}\beta_1\right) = o(m).$$

We begin the future by a careful look at A_{2m+1}. The main point here is (ii). A_{2m+1} can produce $\|S'_k A_{2m+1}\|_1$ of size $\beta_{2m+1}(2m+2)$ throughout $2^{2m} + 1 \leq k \leq 2^{2m+1}$, of size $\frac{1}{2}\beta_{2m+1}(2m+2)$ throughout $2^{2m-1} + 1 \leq k \leq 2^{2m}$. Thus $E'_m(A_{2m+1}) \leq \frac{2\beta_{2m+1}(2m+2)}{m}$. Thus the total future effect is

$$E'_m(A_{2m+1} + A_{2m+2} + \cdots) \leq \frac{2\beta_{2m+1}(2m+2) + \beta_{2m+2}(2m+3) + \beta_{2m+3}(\frac{2m+4}{2}) + \ldots}{m}$$

$$= o(m).$$

We now compute the major contribution which is the present. Here we get combination past future type effects. Let $1 \leq k \leq m$.

$$E'_m(A_{m+k}) = \frac{d'_{m+1}(A_{m+k}) + \cdots + d'_{m+k}(A_{m+k}) + \cdots + d'_{2m}(A_{m+k})}{m}$$

$$\leq (m+k+1)\ \beta_{m+k}\ \frac{(\frac{1}{2})^{k-2} + \ldots + \frac{1}{2} + 1 + 2 + 1 + \frac{1}{\sqrt{2}} + \cdots + (\frac{1}{\sqrt{2}})^{m-k}}{m}$$

$$\leq 14\ \beta_{m+k}\ .$$

Reference

[G] Garnett, J., Bounded Analytic Functions, Academic Press, 1981.

California Institute of Technology

Pasadena, California 91125

Vol. 845: A. Tannenbaum, Invariance and System Theory: Algebraic and Geometric Aspects. X, 161 pages. 1981.

Vol. 846: Ordinary and Partial Differential Equations, Proceedings. Edited by W. N. Everitt and B. D. Sleeman. XIV, 384 pages. 1981.

Vol. 847: U. Koschorke, Vector Fields and Other Vector Bundle Morphisms – A Singularity Approach. IV, 304 pages. 1981.

Vol. 848: Algebra, Carbondale 1980. Proceedings. Ed. by R. K. Amayo. VI, 298 pages. 1981.

Vol. 849: P. Major, Multiple Wiener-Itô Integrals. VII, 127 pages. 1981.

Vol. 850: Séminaire de Probabilités XV. 1979/80. Avec table générale des exposés de 1966/67 à 1978/79. Edited by J. Azéma and M. Yor. IV, 704 pages. 1981.

Vol. 851: Stochastic Integrals. Proceedings, 1980. Edited by D. Williams. IX, 540 pages. 1981.

Vol. 852: L. Schwartz, Geometry and Probability in Banach Spaces. X, 101 pages. 1981.

Vol. 853: N. Boboc, G. Bucur, A. Cornea, Order and Convexity in Potential Theory: H-Cones. IV, 286 pages. 1981.

Vol. 854: Algebraic K-Theory. Evanston 1980. Proceedings. Edited by E. M. Friedlander and M. R. Stein. V, 517 pages. 1981.

Vol. 855: Semigroups. Proceedings 1978. Edited by H. Jürgensen, M. Petrich and H. J. Weinert. V, 221 pages. 1981.

Vol. 856: R. Lascar, Propagation des Singularités des Solutions d'Equations Pseudo-Différentielles à Caractéristiques de Multiplicités Variables. VIII, 237 pages. 1981.

Vol. 857: M. Miyanishi. Non-complete Algebraic Surfaces. XVIII, 244 pages. 1981.

Vol. 858: E. A. Coddington, H. S. V. de Snoo: Regular Boundary Value Problems Associated with Pairs of Ordinary Differential Expressions. V, 225 pages. 1981.

Vol. 859: Logic Year 1979–80. Proceedings. Edited by M. Lerman, J. Schmerl and R. Soare. VIII, 326 pages. 1981.

Vol. 860: Probability in Banach Spaces III. Proceedings, 1980. Edited by A. Beck. VI, 329 pages. 1981.

Vol. 861: Analytical Methods in Probability Theory. Proceedings 1980. Edited by D. Dugué, E. Lukacs, V. K. Rohatgi. X, 183 pages. 1981.

Vol. 862: Algebraic Geometry. Proceedings 1980. Edited by A. Libgober and P. Wagreich. V, 281 pages. 1981.

Vol. 863: Processus Aléatoires à Deux Indices. Proceedings, 1980. Edited by H. Korezlioglu, G. Mazziotto and J. Szpirglas. V, 274 pages. 1981.

Vol. 864: Complex Analysis and Spectral Theory. Proceedings, 1979/80. Edited by V. P. Havin and N. K. Nikol'skii. VI, 480 pages. 1981.

Vol. 865: R. W. Bruggeman, Fourier Coefficients of Automorphic Forms. III, 201 pages. 1981.

Vol. 866: J.-M. Bismut, Mécanique Aléatoire. XVI, 563 pages. 1981.

Vol. 867: Séminaire d'Algèbre Paul Dubreil et Marie-Paule Malliavin. Proceedings, 1980. Edited by M.-P. Malliavin. V, 476 pages. 1981.

Vol. 868: Surfaces Algébriques. Proceedings 1976–78. Edited by J. Giraud, L. Illusie et M. Raynaud. V, 314 pages. 1981.

Vol. 869: A. V. Zelevinsky, Representations of Finite Classical Groups. IV, 184 pages. 1981.

Vol. 870: Shape Theory and Geometric Topology. Proceedings, 1981. Edited by S. Mardešić and J. Segal. V, 265 pages. 1981.

Vol. 871: Continuous Lattices. Proceedings, 1979. Edited by B. Banaschewski and R.-E. Hoffmann. X, 413 pages. 1981.

Vol. 872: Set Theory and Model Theory. Proceedings, 1979. Edited by R. B. Jensen and A. Prestel. V, 174 pages. 1981.

Vol. 873: Constructive Mathematics, Proceedings, 1980. Edited by F. Richman. VII, 347 pages. 1981.

Vol. 874: Abelian Group Theory. Proceedings, 1981. Edited by R. Göbel and E. Walker. XXI, 447 pages. 1981.

Vol. 875: H. Zieschang, Finite Groups of Mapping Classes of Surfaces. VIII, 340 pages. 1981.

Vol. 876: J. P. Bickel, N. El Karoui and M. Yor. Ecole d'Eté de Probabilités de Saint-Flour IX – 1979. Edited by P. L. Hennequin. XI, 280 pages. 1981.

Vol. 877: J. Erven, B.-J. Falkowski, Low Order Cohomology and Applications. VI, 126 pages. 1981.

Vol. 878: Numerical Solution of Nonlinear Equations. Proceedings, 1980. Edited by E. L. Allgower, K. Glashoff, and H.-O. Peitgen. XIV, 440 pages. 1981.

Vol. 879: V. V. Sazonov, Normal Approximation – Some Recent Advances. VII, 105 pages. 1981.

Vol. 880: Non Commutative Harmonic Analysis and Lie Groups. Proceedings, 1980. Edited by J. Carmona and M. Vergne. IV, 553 pages. 1981.

Vol. 881: R. Lutz, M. Goze, Nonstandard Analysis. XIV, 261 pages. 1981.

Vol. 882: Integral Representations and Applications. Proceedings, 1980. Edited by K. Roggenkamp. XII, 479 pages. 1981.

Vol. 883: Cylindric Set Algebras. By L. Henkin, J. D. Monk, A. Tarski, H. Andréka, and I. Németi. VII, 323 pages. 1981.

Vol. 884: Combinatorial Mathematics VIII. Proceedings, 1980. Edited by K. L. McAvaney. XIII, 359 pages. 1981.

Vol. 885: Combinatorics and Graph Theory. Edited by S. B. Rao. Proceedings, 1980. VII, 500 pages. 1981.

Vol. 886: Fixed Point Theory. Proceedings, 1980. Edited by E. Fadell and G. Fournier. XII, 511 pages. 1981.

Vol. 887: F. van Oystaeyen, A. Verschoren, Non-commutative Algebraic Geometry, VI, 404 pages. 1981.

Vol. 888: Padé Approximation and its Applications. Proceedings, 1980. Edited by M. G. de Bruin and H. van Rossum. VI, 383 pages. 1981.

Vol. 889: J. Bourgain, New Classes of $\mathcal{L}^p$-Spaces. V, 143 pages. 1981.

Vol. 890: Model Theory and Arithmetic. Proceedings, 1979/80. Edited by C. Berline, K. McAloon, and J.-P. Ressayre. VI, 306 pages. 1981.

Vol. 891: Logic Symposia, Hakone, 1979, 1980. Proceedings, 1979, 1980. Edited by G. H. Müller, G. Takeuti, and T. Tugué. XI, 394 pages. 1981.

Vol. 892: H. Cajar, Billingsley Dimension in Probability Spaces. III, 106 pages. 1981.

Vol. 893: Geometries and Groups. Proceedings. Edited by M. Aigner and D. Jungnickel. X, 250 pages. 1981.

Vol. 894: Geometry Symposium. Utrecht 1980, Proceedings. Edited by E. Looijenga, D. Siersma, and F. Takens. V, 153 pages. 1981.

Vol. 895: J.A. Hillman, Alexander Ideals of Links. V, 178 pages. 1981.

Vol. 896: B. Angéniol, Familles de Cycles Algébriques – Schéma de Chow. VI, 140 pages. 1981.

Vol. 897: W. Buchholz, S. Feferman, W. Pohlers, W. Sieg, Iterated Inductive Definitions and Subsystems of Analysis: Recent Proof-Theoretical Studies. V, 383 pages. 1981.

Vol. 898: Dynamical Systems and Turbulence, Warwick, 1980. Proceedings. Edited by D. Rand and L.-S. Young. VI, 390 pages. 1981.

Vol. 899: Analytic Number Theory. Proceedings, 1980. Edited by M.I. Knopp. X, 478 pages. 1981.

Vol. 900: P. Deligne, J. S. Milne, A. Ogus, and K.-Y. Shih, Hodge Cycles, Motives, and Shimura Varieties. V, 414 pages. 1982.

Vol. 901: Séminaire Bourbaki vol. 1980/81 Exposés 561–578. III, 299 pages. 1981.

Vol. 902: F. Dumortier, P.R. Rodrigues, and R. Roussarie, Germs of Diffeomorphisms in the Plane. IV, 197 pages. 1981.

Vol. 903: Representations of Algebras. Proceedings, 1980. Edited by M. Auslander and E. Lluis. XV, 371 pages. 1981.

Vol. 904: K. Donner, Extension of Positive Operators and Korovkin Theorems. XII, 182 pages. 1982.

Vol. 905: Differential Geometric Methods in Mathematical Physics. Proceedings, 1980. Edited by H.-D. Doebner, S.J. Andersson, and H.R. Petry. VI, 309 pages. 1982.

Vol. 906: Séminaire de Théorie du Potentiel, Paris, No. 6. Proceedings. Edité par F. Hirsch et G. Mokobodzki. IV, 328 pages. 1982.

Vol. 907: P. Schenzel, Dualisierende Komplexe in der lokalen Algebra und Buchsbaum-Ringe. VII, 161 Seiten. 1982.

Vol. 908: Harmonic Analysis. Proceedings, 1981. Edited by F. Ricci and G. Weiss. V, 325 pages. 1982.

Vol. 909: Numerical Analysis. Proceedings, 1981. Edited by J.P. Hennart. VII, 247 pages. 1982.

Vol. 910: S.S. Abhyankar, Weighted Expansions for Canonical Desingularization. VII, 236 pages. 1982.

Vol. 911: O.G. Jørsboe, L. Mejlbro, The Carleson-Hunt Theorem on Fourier Series. IV, 123 pages. 1982.

Vol. 912: Numerical Analysis. Proceedings, 1981. Edited by G. A. Watson. XIII, 245 pages. 1982.

Vol. 913: O. Tammi, Extremum Problems for Bounded Univalent Functions II. VI, 168 pages. 1982.

Vol. 914: M. L. Warshauer, The Witt Group of Degree k Maps and Asymmetric Inner Product Spaces. IV, 269 pages. 1982.

Vol. 915: Categorical Aspects of Topology and Analysis. Proceedings, 1981. Edited by B. Banaschewski. XI, 385 pages. 1982.

Vol. 916: K.-U. Grusa, Zweidimensionale, interpolierende Lg-Splines und ihre Anwendungen. VIII, 238 Seiten. 1982.

Vol. 917: Brauer Groups in Ring Theory and Algebraic Geometry. Proceedings, 1981. Edited by F. van Oystaeyen and A. Verschoren. VIII, 300 pages. 1982.

Vol. 918: Z. Semadeni, Schauder Bases in Banach Spaces of Continuous Functions. V, 136 pages. 1982.

Vol. 919: Séminaire Pierre Lelong – Henri Skoda (Analyse) Années 1980/81 et Colloque de Wimereux, Mai 1981. Proceedings. Edité par P. Lelong et H. Skoda. VII, 383 pages. 1982.

Vol. 920: Séminaire de Probabilités XVI, 1980/81. Proceedings. Edité par J. Azéma et M. Yor. V, 622 pages. 1982.

Vol. 921: Séminaire de Probabilités XVI, 1980/81. Supplément: Géométrie Différentielle Stochastique. Proceedings. Edité par J. Azéma et M. Yor. III, 285 pages. 1982.

Vol. 922: B. Dacorogna, Weak Continuity and Weak Lower Semicontinuity of Non-Linear Functionals. V, 120 pages. 1982.

Vol. 923: Functional Analysis in Markov Processes. Proceedings, 1981. Edited by M. Fukushima. V, 307 pages. 1982.

Vol. 924: Séminaire d'Algèbre Paul Dubreil et Marie-Paule Malliavin. Proceedings, 1981. Edité par M.-P. Malliavin. V, 461 pages. 1982.

Vol. 925: The Riemann Problem, Complete Integrability and Arithmetic Applications. Proceedings, 1979-1980. Edited by D. Chudnovsky and G. Chudnovsky. VI, 373 pages. 1982.

Vol. 926: Geometric Techniques in Gauge Theories. Proceedings, 1981. Edited by R. Martini and E.M. de Jager. IX, 219 pages. 1982.

Vol. 927: Y. Z. Flicker, The Trace Formula and Base Change for GL (3). XII, 204 pages. 1982.

Vol. 928: Probability Measures on Groups. Proceedings 1981. Edited by H. Heyer. X, 477 pages. 1982.

Vol. 929: Ecole d'Eté de Probabilités de Saint-Flour X – 1980. Proceedings, 1980. Edited by P.L. Hennequin. X, 313 pages. 1982.

Vol. 930: P. Berthelot, L. Breen, et W. Messing, Théorie de Dieudonné Cristalline II. XI, 261 pages. 1982.

Vol. 931: D.M. Arnold, Finite Rank Torsion Free Abelian Groups and Rings. VII, 191 pages. 1982.

Vol. 932: Analytic Theory of Continued Fractions. Proceedings, 1981. Edited by W.B. Jones, W.J. Thron, and H. Waadeland. VI, 240 pages. 1982.

Vol. 933: Lie Algebras and Related Topics. Proceedings, 1981. Edited by D. Winter. VI, 236 pages. 1982.

Vol. 934: M. Sakai, Quadrature Domains. IV, 133 pages. 1982.

Vol. 935: R. Sot, Simple Morphisms in Algebraic Geometry. IV, 146 pages. 1982.

Vol. 936: S.M. Khaleelulla, Counterexamples in Topological Vector Spaces. XXI, 179 pages. 1982.

Vol. 937: E. Combet, Intégrales Exponentielles. VIII, 114 pages. 1982.

Vol. 938: Number Theory. Proceedings, 1981. Edited by K. Alladi. IX, 177 pages. 1982.

Vol. 939: Martingale Theory in Harmonic Analysis and Banach Spaces. Proceedings, 1981. Edited by J.-A. Chao and W.A. Woyczyński. VIII, 225 pages. 1982.

Vol. 940: S. Shelah, Proper Forcing. XXIX, 496 pages. 1982.

Vol. 941: A. Legrand, Homotopie des Espaces de Sections. VII, 132 pages. 1982.

Vol. 942: Theory and Applications of Singular Perturbations. Proceedings, 1981. Edited by W. Eckhaus and E.M. de Jager. V, 363 pages. 1982.

Vol. 943: V. Ancona, G. Tomassini, Modifications Analytiques. IV, 120 pages. 1982.

Vol. 944: Representations of Algebras. Workshop Proceedings, 1980. Edited by M. Auslander and E. Lluis. V, 258 pages. 1982.

Vol. 945: Measure Theory. Oberwolfach 1981, Proceedings. Edited by D. Kölzow and D. Maharam-Stone. XV, 431 pages. 1982.

Vol. 946: N. Spaltenstein, Classes Unipotentes et Sous-groupes de Borel. IX, 259 pages. 1982.

Vol. 947: Algebraic Threefolds. Proceedings, 1981. Edited by A. Conte. VII, 315 pages. 1982.

Vol. 948: Functional Analysis. Proceedings, 1981. Edited by D. Butković, H. Kraljević, and S. Kurepa. X, 239 pages. 1982.

Vol. 949: Harmonic Maps. Proceedings, 1980. Edited by R.J. Knill, M. Kalka and H.C.J. Sealey. V, 158 pages. 1982.

Vol. 950: Complex Analysis. Proceedings, 1980. Edited by J. Eells. IV, 428 pages. 1982.

Vol. 951: Advances in Non-Commutative Ring Theory. Proceedings, 1981. Edited by P.J. Fleury. V, 142 pages. 1982.

Vol. 952: Combinatorial Mathematics IX. Proceedings, 1981. Edited by E. Billington, S. Oates-Williams, and A.P. Street. XI, 443 pages. 1982.

Vol. 953: Iterative Solution of Nonlinear Systems of Equations. Proceedings, 1982. Edited by R. Ansorge, Th. Meis, and W. Törnig. VII, 202 pages. 1982.